STRATEGIC ACTION PLANNING NOW!

A Guide for Setting and Meeting Your Goals

CATE GABLE

*Axioun Communications International
Berkeley, California*

S^t_L

St. Lucie Press
Boca Raton   London   New York   Washington, D.C.

Library of Congress Cataloging-in-Publication Data

Catalog record is available from the Library of Congress   CIP

This book contains information obtained from authentic and highly regarded sources. Reprinted material is quoted with permission, and sources are indicated. A wide variety of references are listed. Reasonable efforts have been made to publish reliable data and information, but the author and the publisher cannot assume responsibility for the validity of all materials or for the consequences of their use.

Neither this book nor any part may be reproduced or transmitted in any form or by any means, electronic or mechanical, including photocopying, microfilming, and recording, or by any information storage or retrieval system, without prior permission in writing from the publisher.

The consent of CRC Press LLC does not extend to copying for general distribution, for promotion, for creating new works, or for resale. Specific permission must be obtained in writing from CRC Press LLC for such copying.

Direct all inquiries to CRC Press LLC, 2000 Corporate Blvd., N.W., Boca Raton, Florida 33431.

Illustrations by Cate Gable
© Cate Gable
St. Lucie Press is an imprint of CRC Press LLC

No claim to original U.S. Government works
International Standard Book Number 1-57444-233-3
Printed in the United States of America   1 2 3 4 5 6 7 8 9 0
Printed on acid-free paper

DEDICATION

For John Hoover, for giving me a chance.

ACKNOWLEDGMENTS

This book could not have been as efficiently written without the emotional support and editorial help of Linda Aldridge and Nancy Jessup.

I would also like to thank my editor, Drew Gierman, at St. Lucie Press, a Division of CRC Press LLC, for his personal interest in the ideas of this book; Barbara Ley Toffler for her insight and conversations about ethics; Stanley Mason for his puckish creativity; and Mark Wilson for his straight-forward assistance.

Additionally, I want to mention a few special people who have helped me at critical points in my career: Edson Sheppard; Rondi Gilbert; Mark Tanaka; Mary Ellen; Susan Harris; Mary Alice Stadum; Ann Golseth; Peter Blomerley; Janathin Miller; Kris Schaeffer; Joe Parker, Joe Profetta, and Mike Ripple; Ninamary Langsdale; and Adele Horowitz.

And, last, there are three people without whom this book could never have been written—William Franklin, Virginia Catherine, and Starla Rae.

THE AUTHOR

Cate Gable is a writer, poet, and business consultant. Founder and President of Axioun Communications International, Gable specializes in strategic marketing and planning for clients both in the U.S. and Europe. Axioun's clientele includes high-tech start-ups, Fortune 500 companies, nonprofits, and small businesses.

Prior to founding Axioun, Gable was Marketing Director for the Federal Reserve Bank of San Francisco and Marketing Manager at CitiBank, Citicorp Western Division. Additionally, she has served as Director of Public Information both for Ohlone College, an institution of 10,000 students in the Silicon Valley, Fremont, California, and for the Federal Reserve Bank of Los Angeles.

Gable has published widely in a variety of fields.

CONTENTS

SECTION II: The Planning Process

LIST OF FIGURES

INTRODUCTION

Ends and beginnings—there are no such things.
There are only middles.

Robert Frost[1]

BEGINNINGS

As a young English teacher on the island of Hawaii, I often gave in-class writing assignments to my students, many of whom were native Japanese studying in Hawaii as a transition to college on the U.S. mainland. One of my favorite students was a shy girl of twelve who talked eagerly outside of class but, when it came to writing, suddenly became terrified. She would sit and look at the blank piece of paper on her desk as if it contained an invisible death sentence. There was something about that blank page and the idea of beginning anything that felt so permanent and precious to her that she could not bring herself to add a mark for fear that she would ruin something. No amount of coaxing could convince her that she knew anything worth putting on that paper.

One day I said to her, "Take out a second piece of paper and put —2— on the top. Now start here, start right in the middle."

Somehow this broke the spell for her and she was able, little by little, to improve both her writing skills and her confidence in English. The idea that she could start in the middle, that she was only adding to something already in progress, freed her to begin.

I tell this story to offer an analogy. Your planning process *is* starting on "page two," whether your organization is a start-up heading toward an IPO or a Fortune 500 company with a 50-year history. There is no such thing as a blank slate. There are traditions and constraints that you will need to consider, attitudes about your product and industry that you will need to take into account, and team members you will be working with who already have a history with one another.

And, further, your planning process is beginning in the middle, literally, in that it is surrounded and influenced by other factors. Your responsibility and your authority are probably limited; but whether or not you can make sweeping strategic decisions about the long-term direction of your organization, you can make a difference—here and now.

In the words of Aaron Wildavsky, "Planning may be seen as the ability to control the future consequences of present actions...Its purpose is to make the future different from what it would have been without this intervention...To change the future, one must be able to get people to act differently than they otherwise would."[2] And that will be your challenge as manager of the planning intervention you have been assigned.

So, in this respect, there really is no beginning; every part of your process is connected to aspects of the past as surely as to the future. However, on the other hand, the important thing *is* to begin: to put that first mark on the paper, to write that first planning idea on the whiteboard.

Both of these ideas—the importance of *action* and the *connectedness* and flow of the planning process to everything that has come before and to everything that will come after—are central themes in the planning approach outlined here in *Strategic Action Planning NOW!*

IS THIS BOOK FOR YOU?

No matter where one begins, beginnings offer opportunity—a chance to get it right this time. It is important to devote some attention at the beginning in order to ensure smooth going later on. Every minute spent at the beginning of a project making sure your fundamental thinking is accurate will save you untold hours later.

Strategic planning is a complex process and, as a discipline, has a huge body of literature explaining, supporting, defining, and analyzing it. You will not find everything in this book, although you will find pointers to many other aspects of planning not covered here in detail.

Before going too much further, make sure that *Strategic Action Planning NOW!* is the right book for you.

If you want a text that leads you point by point through the complex and magical process of creating a mission statement; or you want exhaustive case studies of other businesses and their planning processes; or you need guidance for undertaking a complete financial, competitive, or situational analysis—put this book back on the shelf. Well, you may not want to put it back on the shelf immediately—you might want to take a look at the "Selected Readings" in the back to get some good suggestions for other texts that might satisfy those needs.

What you will find here is a book that, in plain English, outlines an intuitive, team-based, tactical planning process that will allow you to work quickly and efficiently. So, if you need a set of clearly sequenced and explained planning steps; if you want tips and pointers to help ensure that your strategic thinking is insightful and your results are on target; if you have never done strategic planning before but you have said to yourself, "Hey, I'm pretty competent. I've gotten this far, it can't be that hard," then, without a doubt, this book is for you. Read on.

Inherent in the idea of strategic action, as I conceive it, is a commitment to understanding—yourself, others, and the situation in which you are operating. Later chapters will provide further discussion on this; for now, a few words about your work style.

If you are one of those people who never reads the instructions when you assemble furniture, even if it means that the headboard might go on backward and have to be unscrewed and reattached (in fact, you have probably skipped the introduction), you will want to go directly to Chapter 4: Strategic Action Planning: Four Steps. This chapter will give an overview of the four steps of Cate Gable's Strategic Method CGSM© in a quick synopsis fashion. Then, each of the four steps is discussed in more detail in the following chapters (Chapters 5 through 9).

If you are in the "read the directions" category of learner and doer, you would probably rather sneak up on your task, take a few notes, and do a bit of mental warm-up before you begin. In that case, continue from here for an outline of what to expect in each of the chapters of this book.

REVIEW OF BOOK CONTENTS

Chapter 1: My Team, My Self

Strategic planning is a complex process of gathering information and making decisions, and when undertaking a planning project with a team, the process itself becomes even more critical. Chapter 1 offers a discussion about the process-orientation of our planning approach.

The chapter discusses aspects of the team environment: how to balance your role as both the team leader and a planning participant; what aspects of a team-based planning process you can influence and how; tips on choosing team members; a discussion of different models for working in a group; and what to do if interpersonal communications problems arise.

Chapter 2: Beginning Concepts

Chapter 2 proposes some definitions for *strategy* and *action* as single-word ideas and discusses their conceptual power when they are brought together as *strategic action*.

Along the way, the chapter reviews the basic concepts that provide the foundation for the planning process outlined by this book. This chapter might also be of some interest to those who wonder why the Greeks won the Trojan War; that narrative is critiqued as a representative planning process in miniature.

The chapter concludes with a brief interpretation of time, a discussion of the "NOW" in the book title.

Chapter 3: Becoming a Strategist

This chapter explores in more detail the skill of insight, one of the main attributes of a strategic thinker. It discusses how you might actually practice this quality—in the same way you might think of working out in a gym—and enhance your environment to nurture your intuitive strategic skill.

Another important aspect of insight is the ability to be honest with yourself and others. The chapter discusses these ideas using the Asian concept of CHI, meaning energy or spirit, in a metaphoric way applied to the world of business as an acronym for Clarity, Honesty, and Integrity.

Creativity—a key element in any planning process—is also discussed in this chapter.

Chapter 4: Strategic Action Planning: Four Steps

This chapter provides an overview of the entire four-step planning sequence outlined in the book:

- Challenge
- Goal
- Strategic action
- Monitoring and measurement

The following chapters discuss each of these steps individually.

Chapter 5: Seeing the Challenges

It is sometimes difficult to see the things that are right in front of your eyes. This chapter begins the planning process by outlining how to identify or see the challenges in your organization.

Some techniques will be introduced for generating ideas in a formalized exercise called "What Needs Fixing?"

The chapter's second exercise involves a check on alignment between the challenges facing your team and those of your executive leadership group.

In this chapter, hypothetical challenges are created; and subsequent chapters follow them through the remaining steps of the planning process in order to illustrate how each exercise works.

Chapter 6: Sorting the Challenges

In the second phase of the challenge step, you and your team will go through a series of exercises designed to help you analyze and group your challenges into clusters in order to address them more easily.

Chapter 6 also introduces some practical analysis techniques to help in the sorting and prioritizing of your team challenge list.

The sample challenges that started in Chapter 5 will be carried forward, creating a hypothetical discussion to illustrate several of the techniques proposed.

Chapter 7: Setting the Goals

This chapter proposes nine attributes of an effective goal and discusses how to apply them to transform your challenges into a set of preliminary goals.

These goals are evaluated based on some proposed criteria: What is the right amount of stretch? Is the goal reachable and realistic? Is this goal appropriate for your department? Are these goals critical to the success of your overall mission?

An exercise is included to help you and your team with this planning step; included is a hypothetical discussion illustrating how this step might work with the sample list of challenges.

Chapter 8: Devising Strategic Action

Chapter 8 dissects the process of transforming insight and energy into strategic action. Included are discussions about intentional insight and synergy and proposals for some brainstorming techniques to assist your team in formulating innovative action plans.

Additionally, the chapter considers what establishes an environment conducive to creative thinking, and then outlines an exercise that will assist in directing your team's formulation of strategic actions.

Returning to the sample challenges/goals, this chapter reveals their evolution into strategic actions.

Chapter 9: Monitoring and Measurements

This chapter is brought to you by the letter "M," as in monitoring and measurement—the fourth and final step in the planning process. Monitoring is a key factor for success in planning projects, yet it is often overlooked.

Some aspects of effective measurements and a method for adding accountability to your planning process are detailed. There is a discussion about the many ways that monitoring becomes its own tool for planning, management, and even communications effectiveness.

In addition, the monitoring component is added to the sample list of evolving challenges, goals, and strategic actions.

Chapter 10: Implementation Tips

Chapter 10, the post-planning section, provides you with some pointers on how to carry through effectively from the planning stage into the implementation stage of your project.

Various aspects discussed include the planning document, the creation of a communications plan, the formulation of a project implementation team, and an idea for continuous feedback (called the Information and Communications Möbius).

Chapter 11: Future Perfect

As a wrap-up, Chapter 11 suggests a process for self-evaluation of your team's planning process and provides some sample evaluation questions to assist you in improving your planning process the next time around.

Postscript

The Postscript provides some ways to contact the author to let her know if this book has been helpful: what you might suggest changing about it or adding to it; or how you used the book's planning approach with your team.

Chapter Notes

Any footnote in the text indicates further comments on the subject, the quotation citation, or a listing of other resource materials that relate to this topic; these notes appear at the end of each chapter.

Appendices

Appendix A: List of Exercises—including the names of all exercises, what chapters they appear in, and who they are meant for.

Appendix B: Sequential Overview of the CGSM© Planning Process—a visual, chronological representation of the four planning steps and which exercises and tasks relate to those steps.

Appendix C: Sample Planning Items—a sequence of planning items placed together in their natural progression (as discussed in each of the relevant chapters) as challenges, goals, and strategic actions with monitoring devices.

Glossary

Terms used in a particular way in the text are defined in a quick-reference glossary at the back of the book.

Selected Readings

A collection of recommendations for other planning texts you might refer to for further reading or more specific help in areas not covered in this book.

Bibliography

A complete listing of all resources used in writing this book.

Index

An index for topics, businesses, or people referred to or quoted in the text.

So, that is what is assembled for you in this text. The hope is that you will find the book clear and easy to use, full of just the right amount of helpful information, and presented in a way that gives you quick access to what you need.

By the way, studies have shown that whether you read the furniture assembly instructions carefully, sorting, identifying, and counting all the hardware pieces, or whether you begin by immediately attaching part A to part D, the end result is more or less of the same quality and takes more or less the same amount of time. This proves one of the concepts this book's planning process is built on—there is not a right answer, but there probably is a good solution.

Good luck with *your* version of our planning process!

NOTES

1. Lathem, Edward Conney, In the Home Stretch, *The Poetry of Robert Frost*, Holt, Rinehart and Winston, New York, 1969.
2. Wildavsky, Aaron, Does Planning Work, *Public Interest*, Summer 1971, 101.

I

PRE-PLANNING PREPARATION

1

MY TEAM, MY SELF

Organizations must recognize and reward collaboration as clearly and unambiguously as they have traditionally celebrated individual achievement.

Michael Schrage[1]

It sounded like a thrilling planning project.

I was called by the newly-hired, eminently competent CEO of a small start-up with a proven track record and a cutting-edge high-tech product to come in and assist in building a strategic plan with the executive team. All the right phrases were used: participatory, client-centered, clarifying mission and values, vision, process not product, team approach, total quality, continuous improvement, leadership in our industry. I was charmed.

We began meeting—another consultant and I, four division executives, the company founder, and the new CEO—to hammer out a new mission and values statement as preparation for beginning the strategic plan. We filled the whiteboard with words, discussed subtle differences in meanings, argued about and crafted phrases. Then we went on to define and analyze our market, identify our competitors, and talk to analysts about our findings. Finally, we started generating reams of tactical initiatives.

We were joined by another consultant-friend of the CEO who had her own ideas and had not yet been through the ones we had discussed. Some team members had to miss a few meetings to actually "get some work done" with clients. The more words we generated on paper, the more we found that we disagreed about the process itself and the definition of terms. What was a strategy exactly? What was an objective? Who should be in these meetings anyway?

Deadlines were arriving fast and furiously because the planning document was needed for a second round of funding. A writer was called in to format and wordsmith our ideas. The CEO was called out for some high-level venture capital meetings. The leadership vacuum was filled by the consultant-friend who began simply trying to get the job done. There were rumors that we might take over a competitor. Or change our name. One of the planning team members was fired.

In short, our planning project succumbed to the ordinary and customary idiosyncrasies of the real world of business. A mission statement and a planning document were written and the implementation begun, but the emotional and professional fallout was substantial.

What had we done wrong? Or, an easier question, what had we done right? And, more importantly for you, what can you do right before your planning process even begins?

WHAT IS YOUR PLANNING PROCESS?

The most difficult question to ask about your planning project, but the one you must start with, is, "What is our process?" Start by considering some related questions:

- In the context of this planning process, what is your personal decision-making power? Do you know where the outer limits of your authority lie?
- Who will be on your team? Who *should* be on your team? How much leeway will you have to choose?
- What is the territory of your team's decision-making power and the limit of your collective authority?
- How strong will the commitment be from the team members who will participate with you in this planning project? Will they be volunteers or recruits?
- What is the duration of your planning project? How much time do you have, really, to complete the planning phase of the project?
- Who will implement and track the project plan?
- What types of meetings and meeting locations will be needed? Is everyone in one location? Will some of the members of your team be on the phone?
- Do you have funds to use for your planning process? Is an off-site meeting possible? Can you or should you hire a facilitator?
- Is your planning project being mandated from the top down? How far does support for this project extend out into the organization?

- Are you planning in your division simply to meet mandated corporate objectives? Or is this a corporate-wide effort?
- How will you communicate progress to your team members? Is there an intranet whiteboard, chatroom, or electronic posting area for announcements or project updates?
- Will you be expected to report your progress to upper management? When and how? To what extent do you need to communicate your efforts to the entire organization?
- Will the results of your planning process need approval from somewhere other than your team? If so, at what stages? At what levels?
- Will you need a formal planning document or simply guidelines for tracking tactical initiatives?
- How much of your "real job" will you be expected to do while you are facilitating the planning process?
- What kind of support—administrative, budgetary, emotional, political—will you need to adequately accomplish the planning project? Do you know? Do you have it?
- What kind of support will other planning team members have from their managers? secretaries? subordinates?

These are all questions about process. Do not feel overwhelmed by this long list. Some answers to these questions for an ideal planning situation will be proposed; but they are offered as questions because the reality of any planning project is that nothing about the business world or your particular project will be ideal. Every manager knows that everything that can go wrong—and even a few things you thought could never go wrong—will go wrong.

So how can you be best prepared for that inevitability? Just remember that there is no right answer, no right way to do things. The answers to these questions will be dictated by your particular project and the culture and values of your organization. The discussion here will help you sort through the information and respond with a planning process that best meets the given demands, a process that is uniquely yours.

WHAT IS A TEAM?

In recent organizational behavior literature, there is a lot of lively discussion about teamwork and collaboration. There is a difference between the two. A team is composed of members who have the group process and project result as their top, or perhaps for a short time, their only priority; whereas a collaboration is a cooperative effort by a group of people working on their own independent parts of a project.[2]

For example, a corporate project group or task force of individuals all working for the same organization, maybe even the same boss, with an explicit agreement about values and project outcomes, can truly be called a team. Members of this team are dependent on one another for the total success of their project. They will likely be held equally responsible for the outcome of their efforts. Teammates are required to work together and perhaps even their compensation will be tied to how well they work together. More than likely, they know each other rather well. The team might be assembled for a specific project and dispersed when it is completed; or perhaps, it will be a permanent function-oriented structure.

On the other hand, a group of architects, a building contractor, their client, the client's site manager, and local building authorities working on an architectural project would be considered as working in collaboration. These people are dependent on one another for a successful outcome, but they each have their own areas of autonomy and authority. There may be some consensual teamwork early in the process when all the collaborators come together to resolve the final project design; but after the general plan is made, their main reasons for interacting will be primarily for task hand-off, timetable coordination, and problem resolution.

Peter F. Drucker, in *Managing in a Time of Great Change*, describes three types of teams based on sports models.[3] The baseball team is more what we might call a collaboration of individuals; these players each hold a separate position on the team and they work together within the fixed requirements of their positions, but they do not necessarily play as a team. In football, team members each have a fixed position but they play more clearly as a team. A tennis doubles team functions even differently; each player is responsible for a territory or primary position—net or baseline, left or right—but must adapt as needed to provide back-up, depending on the partner's moves.

In fact, there are many more unique types of teams than just these three. What about an Olympic ski jump team? There, individual performance is primary, but the team still makes decisions about sequence of jumpers; and the team's total score is what wins. This might look similar to the baseball team model; but unlike the ski jumpers, the ball players' team interactions are still critical to their success.

Or what about the recent Scrabble team of Matt Graham and Joel Sherman, two independent champions joining forces to compete against Maven, the computer program created by Brian Sheppard?[4] One might argue that Graham and Sherman teamed as if they were tennis doubles partners, although the pace and requirements of the game made their calculated interactions and decision-making process unique. There are as many ways to organize a group to solve problems as there are organizations and problems that need to be solved.

But, in any case, what difference does it make to label a group a "team," a "collaboration," a "task force," or an "ad hoc committee?" The name is arbitrary. What is critical is that all of these differing labels reflect functional differences in the working relationships between the group members, their decision-making processes, and their behaviors and commitments to one another. In other words, their group processes are different and each type of work-group structure makes differing requirements of the group leader—if there is one—and its members. Understanding and/or manipulating these differences allows you to be more effective in your role as project manager or team leader.

How a team or collaborative group works together will also differ, depending on what kind of project is being undertaken. If a group of people will be together for a long project—for example, over the course of many years—their work styles together will evolve in a different configuration than if the same people had been called together for a shorter project. If a member of your work team on a longer-term project is having trouble, team members might spend extra time coaching this team member on weekends or after work, giving her the information she needs to get up to speed; whereas, if the group is working on a project with a short time-horizon, there will not be the motivation or payoff for that kind of time investment.

As Drucker points out in *Management*, "the more flexible an organization [or team] is, the stronger do the individual members have to be and the more of the load do they have to carry…and the more clearly a structure defines work, authority, and relationships, the fewer demands does it make on the individual for self-discipline and self-subordination."[5]

What this means for you as team leader is this: although there may be less flexibility in the structure, it will be easier for your team members to work efficiently if you provide clear guidance on the following aspects of your project:

- Team process: how you will be working together, structural definition
- Work objectives: what you are mandated to accomplish
- Values and agreements: your cultural, behavioral environment
- Team roles and responsibilities: who does what, when
- Commitment expectations: duration and scope of the project

These aspects should be open for discussion, but you will need to be ready to propose what you think will work best to get the job done. Tom Peters in his classic, *In Search of Excellence*, offers this advice: "Teams that consist of volunteers, are of limited duration, and set their own goals are usually found to be much more productive that those with the obverse traits."[6]

Moving forward, the following two chapter sections explore some other aspects of the appropriate team dynamic for your planning project, followed by a team exercise at the end of the chapter.

WHO IS YOUR TEAM?

If you have the good fortune of being able to decide who will be on your team, pick the people who will do the best job. Simple. However, not always possible. Sometimes you may not really know who would do the best job. Individuals who possess, by themselves, admirable qualities can sometimes contribute in a negative way to whole team dynamics. Sometimes, for political reasons, your choice will be hampered by the need to include certain people; or you might be expected to appoint a certain level of personnel when you know other individuals who have a special interest in the project and would do a better job. As Peters points out in the quotation above, interest in serving should be one of the main criteria for appointment.

Whatever the case may be, here are a few things to keep in mind:

- Pick your team members from departments, divisions, or functional areas that you know will be needed to accomplish the planning goals. It is very difficult to involve people in the implementation part of the process who have not been included from the beginning.
- Choose individuals with a diversity of opinion and approach. This will make your discussions lively and provide an innovative atmosphere for problem-solving.
- Choose individuals with a range of expertise and knowledge. This will allow you to compose a strong version of what Michael Schrage terms, the "communal mind."
- If all else is equal, choose good communicators. H. Clayton Foushee, a former technical adviser to the FAA, says, "We have done studies of intracockpit communications and the conclusion is that more communication was correlated with better performance."[7] One would guess that the same is true of an effective team.

- Select people who are interested in planning. Some people are just naturally brilliant strategists; others would rather be *doing* the activity than *planning* it.
- Pick teammates who have time to devote to your planning project. If the majority of team members have this planning project as second or third priority, you will not have a very motivated group.
- Do not pick too large or too small a group. My suggestion is that a core group of between 8 and 12 key project members is a good size.[8] There may be exercises that these members take to other colleagues in their respective teams and, as seen in the chapter exercises, the core group will break into small groups from time to time. Keep your core membership manageable. It is very debilitating to have a large planning team, 40% of whom are not present or not interested.

So, if selection preferences can be taken into account at all, that will help you to begin your process on the right foot. However, if this is not possible, do not despair. Your organization is made up of capable and intelligent individuals or they, and you, would not be there at all. If you do not have a choice, take what you are given and make the best of it.

WHO ARE YOU AND WHAT IS YOUR ROLE?

No, there is no need to proceed into an existential discussion of the essence of being, but it is necessary to think about your qualities and your role in this planning process; this will help you tremendously as you proceed.

First of all, it is assumed that you, the reader of this book, are a middle-manager responsible for a team-based planning project in a corporation, nonprofit, government agency, or even a small business. It is also assumed that you will not be completing this planning project alone. You will have a group of colleagues that will work with you to generate ideas and action plans. These core planning team members may also be implementors, or perhaps the supervisors of plan implementors. And you will probably have the dual role of both participant *in* the planning process and facilitator or leader *of* the process.

This dual role carries with it some advantages and some disadvantages. The advantages are that, to the extent you can, you will have more power over the process than the other members of the team. You can suggest guidelines for the process itself. You will, whether you are conscious of this or not, set the tone for the quality of the debate and the thinking. You will be the one with the most responsibility and, therefore, the one with the most to gain if the project goes well.

On the other hand, people will blame you if things go poorly. And, as a participant and a facilitator, you will often be torn between these two roles, not knowing which to follow. Should you put your opinion out forcefully because you feel strongly that it is the right way to go? Should you intervene and mandate a decision? Or should you hold back and encourage other members of the team to speak their minds? If you are a strong leader, you may be criticized for taking too much power. But if you allow the process of discussion to unfold in a more leisurely way before a decision is reached, you may be accused of not moving the process along fast enough.

I have a friend who is a consultant to high-level executives. In the workshop or team meeting setting, when she wants to make a point as an individual and not as a facilitator, she actually takes two steps sideways and says, "This is Susan, the person, speaking now. I think..."

How you mediate between these two roles you will play—as team leader and team member—will, to some degree, define what you feel it is to work as a team and in a team. No one is going to answer this question for you, although people may have opinions about it. It will be up to you to find a stance, a balance, a rhythm that feels right for you.

I was the foreman of the jury in a murder trial several years ago. It was a trial that brought up strong feelings in all of us jurors because of the personalities involved and the graphic nature of the crime. In fact, after the first day of testimony, one of the jurors did not return because she was too upset by the content of what she had heard.

Selecting the foreman of a jury is the first job the jury does as a group after it is sent to begin deliberations. At that point, everyone has sat through weeks of testimony, without the benefit of being able to talk to anyone else about it, and is bubbling with questions, opinions, and theories. One craves the opportunity to discuss with one's peers, to puzzle through something seemingly so incomprehensible in order to make sense of it. And it is just at the brink of this rich, conflicted discussion that the group must make a decision about who will lead the process of discussion.

When I was chosen as the foreman, I felt two things simultaneously. First, that I had been given an honor that I was proud to accept and would carry out to the best of my ability. And, second, that in exchange, I would have to partially give up my right as a jury member to participate fully in the vehement discussion I knew was bound to take place. That was my solution to the situation. Nobody said it had to be done that way. But it was my way of dealing with two roles that I felt held some inherent conflicts.

You will want to think through what it means to both lead and participate in your own planning process. What boundaries will you draw between the two roles you will play?

WHAT IS YOUR TEAM CULTURE?

One of the important tasks that both you and your teammates will need to accomplish very early in your project is to establish what the environment for your process will be. By that I mean the values and agreements that will be in place as you work together.

Your planning project exists within the larger context of the culture of your organization, but that does not necessarily mean that the culture or environment that you create for your project will be a replica of what exists in your company. Your planning project will have some different requirements, depending on its nature; and, as the leader, you have a right to create the culture that will best fit your project.

What is strongly recommended is that you bring your team members together, before the actual planning agenda activities are started, to discuss and agree on the playing rules and the values you all feel are important in the process.

These rules could be as tangible as some assumptions about attendance and timely arrival for meetings; or as fuzzy as the establishment of the ways you and your teammates agree to disagree.

The planning process that is outlined in this book assumes certain attitudes or values for your teamwork. Simply stated, your credo might include the following tenets as positive attributes of an effective planning environment:

- Consensus-oriented decision-making
- A respect for individual contribution
- A belief in the strength of the synthesis of ideas and, therefore, a value in diversity of thought
- A belief in the power of innovation and synergy
- A need for self-awareness in team members
- A commitment to honesty, integrity, and kindness
- (What we called in one organization where I worked) an agreement that there will be "no body slams"

Your team might want to come up with its own set of rules or principles that will establish an environment for your discussion. But whether you use some of the ideas suggested here or devise your own set, having the discussion with your team and coming to your own agreement about values is crucial. Even if these values are already strongly reinforced within your organization, it is always a good reminder for teammates to hear them again. And if these are not values that are a part of your existing corporate culture, there will be even more need for your establishment of clear rules of conduct for your working group.

The bottom line is that the meeting environment for your team be considered a safe place for discussion and one that is conducive to creativity and collaboration. These aspects are discussed in more detail as we progress through our planning steps.

Here is an exercise that should be conducted at the first meeting of your core planning team. It outlines a process "to plan to plan." This plan to plan could be the most important agreement you make as a group.

EXERCISE 1: CREATING AGREEMENT

Materials: flip chart and marking pens.

Duration: 60–75 minutes. If the meeting goes longer than this, schedule part two for a different time.

Objective: to come to an agreement about the following aspects of your planning process:

1. Discussion of the process
   Review of the CGSM© Four-step Planning Process and Definitions
   How to hand-off plan to implementation team
   Calendar of meetings and project timelines
2. Discussion of roles and responsibilities
   Introduction of core group members
   Team leader: introduction, responsibilities
   Others: support functions, implementation team, etc.
3. Discussion of values and agreement about rules for behavior at meetings
4. Group statement of commitment: could be a team mini-mission and values statement

Facilitation tip: This will be a whole group discussion and the first time the team has met as a group, so you may want to start with some kind of warm-up exercise. A good one is the "interview someone else in the room and introduce them" warm-up. For a variation on this: in the introduction, include one false thing about the person you are introducing and have the group guess what it is. Another introduction exercise is to have each member bring and read a favorite passage from a poem, essay, play, or novel and talk about why it is important. Or another: ask team members to bring an object or photo from their childhood and talk about its significance.

This entire first meeting will be a whole group discussion interspersed with your suggestions for or explanation of certain parts of the agenda. At each point outlined above, be sure that the group has reached agreement before continuing. Sometimes, it is a good idea to actually write up the points agreed to and send them out as the first team communication.

WHAT IF?

Now, what could possibly go wrong within your team framework? We are not talking here about all those outside influences that will impinge on your project—i.e., other work deadlines, industry turmoil, corporate pressures, a bold announcement of a new technology from one of your competitors, the flu—but those ways that team members can get into difficult situations with one another.

As the leader of your team, you probably possess a broad range of honed and tested people skills or you would not have been selected to take on this task. But that does not mean that you may not want to spend a little time thinking about some contingencies if team personality disturbances erupt and threaten to destroy your project.

No doubt, despite the competency of your team members, you will get on each other's nerves from time to time. A tongue-in-cheek compilation of the types of people that can play havoc with even a generally compatible group of colleagues is given below:

- The doubting Thomas
- The side-tracker
- The cute-but-irrelevant-comment maker
- The dominator
- The smirking, silent type
- The inappropriate joke teller
- The PC advocate
- The offer-no-ideas-but-criticize-the-results skeptic
- The everything-is-perfect optimist

You may recognize some of these and, no doubt, be able to add many more to this list.

The problem is what to do to counteract the tendencies of those members whose contributions or whose style of communication look as if they could sabotage team progress. Your establishment of the ground rules will help; but in tense situations, everyone has the capacity to revert to unproductive or abrasive behaviors.

A few simple suggestions:

- During the meeting, deal with the problem immediately and directly—first by gently reminding the offending member of the stated meeting values/rules. Depending on the severity of the offense, you may want to follow up with a private discussion after the meeting.

- If the problem persists, talk to the individual with the individual's supervisor or your company's human resources director present; and/or, as a last resort, find a replacement member for your team.
- Sometimes, behavior that seems charming to one person is offensive to another because of cultural differences. Personal space, ways of touching, gestures, tone of voice, accent or inflection, and styles of communication can vary tremendously from culture to culture. If this is the root of the problem, you might need to seek out an expert to assist with some mediation or cultural diversity education for your team.[9]
- If two team members are engaged in an overly heated and disruptive discussion, suggest a time-out for the group. Then take the two individuals aside for a private discussion and attempt to resolve their differences.
- Do try to make sure that all members of your team have the chance to contribute to the discussion. If someone seems to be dominating the discussion, feel free to turn to a silent member and say, "What do you think, Joel? We haven't heard from you yet."
- *All* cultural or ethnic jokes are inappropriate. If someone tells one, perhaps the best strategy is to go on as if nothing had happened and take that person aside later. This sends a message of disapproval without creating a disruptive or confrontational situation in the meeting.
- Abrasive behavior can indicate that the individual is simply uncomfortable with something: another team member, an inability to participate in the discussion for some reason, a problem that exists outside of work. This is not to suggest that you become a therapist, but there may be times when a frank and private discussion with the individual in order to coax out the underlying issue might be helpful.

For some particularly cogent analyses of meeting discussions, take a look at *Talking from 9 to 5*, Deborah Tannen's excellent book on communications in the workplace.[10] Its focus is "How men's and women's conversational styles affect who gets heard, who gets credit and what gets done at work," but the principles discussed are relevant in any kind of team setting. Of particular interest for the application here is Chapter 9, "Who Gets Heard?: Talking at Meetings."

SUMMARY

Thinking about the kind of planning project you are anticipating, the length of its commitment, who might be involved, and how you want to interact with your teammates may help you answer some of the questions about your group process. In a self-reflective way, this will in turn clarify for you what kind of working structure will be best suited for your planning project.

Chapter 1 focuses on team process because your team or collaborative work group will be the foundational "home and family" for you and your planning project for the upcoming months. Much of the success of your project will be based on your team, its make-up and dynamics. So in focusing first on your team and yourself, you are creating the nesting ground for your project.

As with so much of the planning process that this book will assist you through, many of the aspects of this team—its working rules and environment, and your role in it—will need to be constructed from the "givens" around you. If the question is, "what kind of work group do we need for our planning project?"—then the right answer is "whatever structure enables people to perform and contribute" to the best of their abilities.[11]

Each team and planning process is the unique product of a particular corporate environment based on constraints, circumstances, personalities, chance, hard work, competency, and a hundred other factors special to you, your teammates, and your business. And that *is* thrilling.

NOTES

1. Schrage, Michael, *No More Teams! Mastering the Dynamics of Creative Collaboration*, Currency, Doubleday, New York, 1995, xiv. I highly recommend this book. Although the incendiary title suggests that teamwork is now passé, in fact, Schrage is simply redefining techniques for effective collaboration.
2. My discussion is a slight departure from Peter F. Drucker's definition in *Management: Tasks, Responsibilities, Practices*, Perennial Library, Harper & Row, New York, 1974. In this publication, Drucker defines a team as "a number of people—usually fairly small—with different backgrounds, skills, knowledge, and drawn from various areas of the organization (their "home") who work together on a specific and defined task" (p. 564). For more detail on work groups, take a look at the entire Chapter 45, "Work- and Task-Focused Design: Functional Structure and Team."
3. Drucker, Peter F., *Managing in a Time of Great Change*, Truman Talley Books, Dutton, New York, 1995, 97. The ideas I summarize are taken from Chapter 8, entitled "Three Types of Teams."
4. Information is taken from the *New York Times Magazine*, May 24, 1998, an article entitled "Humankind Battles for Scrabble Supremacy," p. 20–23. What's interesting here is that Graham and Sherman, both Scrabble champions in their own right, have very different styles of play yet needed to negotiate with one another to come up with moves approved by both. Their strategic thinking in one critical move, which would not have been conceptually possible for Maven, proved that man's ability to anticipate behavioral patterns and act on those intuitions is still superior to the rote memory of Maven's program.

5. Drucker, *Management*.

6. Peters, Thomas J. and Robert H. Waterman, Jr., *In Search of Excellence*, Time Warner Books, New York, 1982, 129.

7. H. Clayton Foushee's comment is quoted in Schrage's book, p. 12 (see Note 1).

8. Peters, Thomas J. and Robert H. Waterman, Jr., *In Search of Excellence*, Time Warner Books, New York, 1982, says ten or fewer (p. 129) make a task force and notes that academic evidence indicates the optimal number of individuals is about seven; but this is a definition for 'small group' (p. 127). Peters also has some discussion about team structure in Chapter 5, "A Bias for Action," p. 119–155.

9. There are many kinds of cultural diversity materials available. A simple work-book-type approach is provided in *Working Together*, by Dr. George Simons, Crisp Publications Inc., Menlo Park, CA, 1989.

10. Tannen, Deborah, *Talking from 9 to 5: How Men's And Women's Conversational Styles Affect Who Gets Heard, Who Gets Credit and What Gets Done at Work*, William Morrow and Company, New York, 1994.

11. Drucker, *Management*, 528.

2

BEGINNING CONCEPTS

*The nice thing about not planning is that failure comes as a
total surprise and is not preceded by a period of worry or
depression.*

Malcolm McDonald[1]

SO WHY PLAN? WHY WORRY?

Sometimes, success can sneak up on you just the way failure can; blind
luck, common sense, good timing, or even a sense of direction can make
the difference.

Take, for example, Robert L. Page, owner of Replacements Ltd.,[2] who,
without one word of a business plan, built a seemingly wacky idea into
a $57.5 million business in 16 years. What does he sell? Old tea cups,
saucers, and plates—one tea cup at a time.

Mr. Page stocks 47,000 china patterns, many of which went out of
production years ago. His motto is, "We Replace the Irreplaceable" and
his main claim to business fame is his systematizing the stuff of flea
markets. His highly computerized business tracks both customer requests
and the Replacements Ltd. inventory and spews out over 50,000 pieces
of mail daily. One of the 144 phones lines is always ringing.

Mr. Page says of his success, "The basics I thought of, but the growth
is something I never considered. I never dreamed of being a millionaire."

Of course, he may want to take to heart that old Chinese proverb:
"The first one hundred years are the hardest."

Good luck and a flare for marketing might get a business established,
but it is only insightful planning that ensures sustainable success. In the
beginning, a business owner knows what is going on everywhere in her

business because she is doing most of the work herself. But the larger an organization grows and the more fragmented the organizational structure becomes, the greater the need for a coordinated planning effort. But even with a strong strategic plan, some businesses get caught off guard when major aspects of the environment or industry change. Upcoming chapters will include some of these business scenarios.

WHERE DO YOU FIT IN YOUR COMPANY'S PLANNING PROCESS?

Some aspects of your planning process from your team's perspective have already been discussed. Here are a few more ideas on the politics of change within an organization. Any successful planning process begins with the assumption that you and your organization understand the value of planning and that you, as a manager responsible for a planning effort, have the highest level of commitment in support of your efforts.

Since you are reading this book, one assumes that you have been given a planning project by someone in your organization who understands the importance of planning and will give you the support you need. If that is not the case, you will want to go back to your boss and do some bridge-building.

As alluded to, no planning process at any level can succeed in isolation; ideally, your efforts are part of a coordinated planning project happening organization-wide. One of the most common reasons that planning fails is because the process is not supported by key decision-makers at the top. There is no amount of lip service that can replace genuine support. If a CEO or corporate leader wants a planning process because it looks good but has no intention of allowing departments to devise and carry out their own plans, run the other way—or, more prudently, offer a graceful excuse and volunteer to serve on a different task force.

Change in an organization cannot happen unless those in power want it to happen; or until the power structure is so endangered, because of the poor health of an organization, that a new structure emerges. Even change that begins on a grassroots level must ultimately be supported by top management in order to be successful.

There are cases, however, in which the CEO is the change agent; the following is an example.

SEI Investments in Oaks, Pennsylvania, founded in 1968, handles back-office operations for trust departments at 85 of the top 200 U.S. banks and offers an investment advisory service for wealthy individuals.[3] SEI is run by founder Alfred P. West who serves as Chairman and CEO. West became a man with a vision.

In January 1990, West broke his leg in a skiing accident and spent three months out of the office. That gave him some time to mull over concerns he had about his organization's business structure. West found himself thinking back fondly to the early days of his business when there were 60 people working in a big open room with complete access to one another—what one might call "continuous communication" today.

The more West thought about it, the more the current business structure of SEI—the hierarchy, the divisions, the multilevels of senior vice-presidents—began looking cumbersome and lead-footed. When West returned to work, he wanted change; and what he brought with him was the notion of 'fluid leadership.'

Mark Wilson, current member of the corporate team and someone who stayed through the, sometimes traumatic transformation, says there were two waves of reinvention. The first wave involved intensive planning sessions for every division of SEI, but the teams were still functionally centered at that time; and during their planning, they realized that they needed better access to one another for support and information exchange. By the second reinvention phase, everyone had reteamed cross-functionally.

Now, SEI is organized into 140 self-managed teams made up of sales and support, marketing, technical, and account members. There are no prearranged offices or floor plans—nothing to denote title or executive privilege. The office design is strikingly representative of the "fluid leadership" concept—open space with coiled cords dropping from the ceiling to deliver electricity, Internet and phone access—that allows work groups to form or disband depending on business initiatives. As one of the facilities designers commented, "We wanted to throw away any structure that would impede us from doing what was right for the business. Every day, it's a new place."

West also flattened the management hierarchy. There are no secretaries. Managers are required to do their own typing, faxing, and travel planning, which West said, "really broke the back of the old culture." West himself can be seen standing at the copy machine from time to time.

SEI's new corporate culture is innovative, fleet-footed, closer to the customer. They have reinvented themselves. But if employees were not willing to get on the train, they were left at the station. Two of SEI's three division heads quit; a third was fired.

On the other side of the coin is an example of change inhibited by the decision-makers at the top. A friend of mine who worked at a large publicly held industrial firm made suggestions to upper-level managers about changes that she felt were critical if her division was to remain a contributing business unit. Her suggestions were summarily brushed aside and she moved on eventually. She heard later, via the corporate grapevine, that two different sets of consultants had been hired to do research on sagging sales and had recommended roughly the same set of changes that she had outlined. In every case, the suggestions were perfunctorily rejected by management.

Just last year, the Board took the reins, some upper-management heads rolled, and the much-needed changes were initiated.

The moral of these stories?: If top management wants change, it will happen with or without you—make your choice. And if they do not want the changes that you feel are needed, do not wait, make your move.[4]

In fact, most planning procedures start at the top. You may have been mandated to flesh-out a planning process that has begun in the higher reaches of your organization by the top members of the visionary team. Generally, it is at this level of management that the overall strategic direction or strategic vision is established. Something like, "We will establish market leadership for XYZ Inc., by increasing sales in our existing customer base by 8% this year." Or "We will enter international markets B and C with the anticipation of a negative ROI until 20XX." Then this strategic vision is handed down for implementation to various departments and divisions for tactical implementation.

Top-down is not the only model for change, however. Many grassroots changes are possible. At SEI, although West was the visionary catalyst, he sparked a grassroots movement that involved all employees at all levels. The revolution came up to meet his vision.

It is worth making a distinction here between financial planning and strategic planning. Although they support one another—obviously, one needs to consider the budget for what one will be asked to implement—they are different in many respects.

Financial planning is that yearly exercise every department goes through in order to formalize dollar allocations. Generally, this kind of planning assumes a status quo condition. In fact, many companies simply transfer categorical budget figures from one year to the next, occasionally taking into account increases for foreseeable changes, like a system conversion or opening a new office. In other methods, a department must start from zero every year and account for all monies needed rather than budget from a status quo baseline. In any case, this is not *strategic* planning; this is an accounting exercise.

Some planners would even argue that efficiency plans for operational improvements are not strategic planning, under the assumption that "strategic" relates only to the marketplace and to honing the competitive advantage of your organization. In discussing the difference between the strategic and operational planning modes, Kenichi Ohmae, a well-known Japanese strategist, says, "It is the difference between going into battle and going on a diet."[5]

In this process, consider both internal operational efficiencies and external market growth strategies as being part of a whole—activities of either type might become part of a strategic action plan for your unit. Whether you are part of the top management team revising or creating a corporate mission and vision statement, or you are simply carrying out your department's implementation programs in support of a corporate strategy, the steps for the planning process are the same.

WHAT IS STRATEGY?

In preparation for taking the first planning step, one needs to look at a few foundational concepts for the planning process. Start with the phrases used in this book's title: What is "strategy"? What is "action"? And what is "NOW"?

The concept of strategy as it applies to planning has been borrowed from the military, as have so many business metaphors, because contained in the notion of strategic planning is the idea of winning by overcoming a competitor. In fact, the etymological root of strategy is a Greek word *strategos* that means general, or someone who commands a *stratos*, or army. But strategy is not a tool used only in the war room or the board room.

From the time we are born, we begin developing strategies to get what we want. We use whatever resources are available to us or whatever is under our control—however limited—to get what we think we need. Shrieking substitutes for, "I'd like a meal now please," and waving our hands wildly in the air sometimes lands us either something fun to touch or one of those big smiling creatures that make us laugh. Nor is the use of strategy a skill possessed only by humans, as any pet owner knows.

I mention this only to point out that *homo sapiens* are thinkers—we specialize in strategies. In fact, *sapiens* is derived from the Latin verb *sapere*, to taste or to perceive, to be wise. You do not need a special planning course or an advanced degree to become an effective strategist—although training can assist you in perfecting your innate skills. You have what you need right now, in your bank of business and life experiences. This book will simply help you sort things out a bit by outlining a common-sense approach to planning.

When one thinks of strategy, some of the following words or ideas might come to mind.

- Agility
- Precision
- Surprise
- Insightfulness
- Efficiency

- Cleverness
- Imagination
- Innovation
- Using the resources at hand
- Being prepared
- A plan of attack

In a military sense, a strategy or stratagem is a maneuver designed to deceive or surprise an enemy or a clever, sometimes underhanded, scheme for achieving an objective.

In general terms, a strategy is something that drives or governs a set of actions intended to accomplish a specific purpose. A good strategy might be clever, or something no one else has thought of, but it does not necessarily need to be. A strategy does not need to be unique in the world to be effective—it only has to be effective enough to achieve the desired goal. In fact, it is much more important that the strategy be appropriate to the situation than that it be innovative, although often an innovative strategy delivers an advantage.

A strategy should outline actions that utilize the resources available or provide a means of getting needed resources in order to accomplish a specific goal. If the strategy is appropriate to the challenge, if it is insightful, and if the enactment of the strategy is well-executed, the result should be the successful achievement of the intended goal.

Notice that there is an important sequential relationship here. The strategy comes before the action. And the goal is not only the catalyst or reason for conceiving the strategy, but it is also where, if all goes well, the action ends up.

Figure 2.1 outlines the sequential relationship of the main steps of any generalized planning process. Although one will see many different terms used for these steps, particularly as levels of planning detail are added, these steps and their relationships to one another remain the same.

As Figure 2.1 illustrates, the progression of the major planning steps forms a kind of tumbling loop or triangle. In an effective planning process, each step leads to the next and, at the same time, each step is inextricably connected to the previous steps. This connectedness is what gives the loop/triangle its integrity and strength; if any step falls out of its connectedness with the others, the integrity of the loop is broken and the process is stopped. This graphic provides an apt visual metaphor for a good planning process.

Additionally, Figure 2.1 is drawn to indicate that the planning cycle is continuous. The goal is always being revised and updated as business conditions change, which in turn leads to changes in the strategy, action,

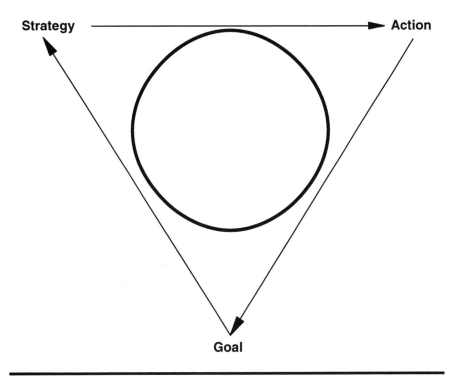

Figure 2.1   Basic Planning Cycle

and then the goal again. Actually, it might be more accurate to draw the schematic as a spiral, like an uncoiled spring, because each successive cycle of planning covers related-but-new ground.

As one progresses through the planning steps outlined in this book, there will be further discussion about the sequential relationship of these basic planning elements and their referential interconnectedness, both to the steps before, and to the steps following. This basic planning structure will be used to discuss, in more detail, the substeps, or subtasks, within each of these major planning milestones.

For now, continue to focus on strategy by looking at some simple strategies in more detail to find out what general attributes make for a good strategy.

WHAT ARE THE QUALITIES OF AN EFFECTIVE STRATEGY?

As an example, say the boys have a clubhouse and will not let the girls in. The boys might put up, with the greatest bravado, a "Do Not Enter" sign. The girls, of course, will ignore them. They will probably put their hands on their hips and go on about their business. Maybe they will make their own clubhouse and nail up their own "Do Not Disturb" sign.

Neither the girls nor the boys will mandate specific actions—OK guys, let's make a sign. OK girls, when the guys put up the "Do Not Enter" sign, put your hands on your hips, raise your chin, and turn to the right. But both groups have outlined a simple strategy—a general concept that governs a series of actions. In this case, both parties are hoping that their actions will have a certain desired effect or goal: that they will get noticed, probably, in this case, without appearing to want to be noticed.

If one extrapolates from this scenario, one might agree that, in simplest terms, a good strategy must be:

- Clear and easy to communicate to others
- Agreed to
- Flexible enough to allow for changing circumstances

Often, a strategy is preceded by a 'situational analysis' that establishes guidelines for the strategy or outlines current conditions. These conditions often indicate something that must be overcome in order to win an advantage. In the nomenclature of this process, these are called 'challenges.'

A simple military example of a situational analysis might be a report that notifies one side that the other side has more resources, either in manpower or fire-power. In this case, the pre-condition of unequal resources—or what is called a challenge in this planning approach—outlines one of the parameters for the strategy in that it is something that must be overcome.

The high command in a military scenario might stipulate, "Since there are fewer of us than them, we've got to make them think that there are more of us." This might translate into the strategic idea, "We've got to surprise them." Then the strategy must include some tactic that will be surprising or will make the enemy think they are outnumbered.

Moving to a business context, one might say, "Our strategy this year is to have the best product at the lowest cost," or "We're going to be known for our exceptional customer service." These strategies might be based on a more thorough competitive or situational analysis that indicates that they are likely to succeed against the competition in the current environment.

In each case, the strategy could lead to many, varied actions depending on several factors—resources, regulations, ingenuity, a country's social and cultural restraints, an organization's corporate culture, the skills of the people involved in executing the action, what the boss wants, etc. *Every* situation is unique.

The Trojan War was a particularly difficult and trying battle for the Greeks. After their 10-year experience with the fierce Trojan warriors, their cumulative "situational analysis" indicated "we can't defeat them without getting inside their well-fortified city walls." So, the Greeks took this challenge into account and held a strategic planning session on the battlefield in front of Troy's Scaean gates.

Odysseus, perhaps the wiliest strategist of all time, came up with a creative, even artful, action based on the strategy of surprise.[6] His idea was the Trojan Horse. The horse was a huge, beautiful, wooden sculpture that would hold their bravest and boldest Greek chieftains inside it. The Greek plan was to appear to retreat to their ships and place the sculpture in front of the Gates of Troy as if it were an offering. Then, by means of some elaborate storytelling, they hoped to convince the Trojans to bring the horse into their city.

Maybe the reader has noticed that the one thing missing from this brilliant strategy is a clear statement of the goal—that concept which should begin and close any strategic action loop. This might seem like an obvious omission but, often, planning projects can progress on their own momentum right up to the point of action without anyone ever considering clearly what the goal *should* be.

Once the Greek warriors jumped out of the giant Trojan Horse in the city of Troy, would they be clear on their motives? Did they want to capture the city, kill as many of the enemy as possible, or re-kidnap Helen? After going to all the trouble of constructing a huge hollow horse (imagine what their procurement officers thought) and getting it into Troy, the Greeks would certainly be in an awkward position if they had to jump out of their device in the middle of enemy territory and discuss what to do next.

This story is used to propose as a working summary that a good strategy provides clear and imaginative guidance for the effective use of available resources in order to reach a specific goal.

Now look at the action step in the planning cycle to see how these two concepts—strategy and action—fit together.

WHAT IS ACTION?

Strategy exists in the world of ether, in ephemeral grey matter, in that place about which scientists and theologians have been debating for centuries: How many strategies can exist on the head of a pin? Does a thought have metaphysical substance? Perhaps. But for marketing and sales purposes, not yet. A strategy is a thought, a plan for action, but a plan only. It has no effect in the world until it is implemented.

Until one puts a strategy up on its feet as action, no goal can be reached. One can see, feel, or hear action. Action is of this physical world—it involves the direct manipulation of the world and its objects. Thus, the skillful implementation of a strategy is the natural and only useful partner to strategic insight. Chapter 9 offers some tips on the implementation phase of the planning project.

When one thinks of action, some of these words or ideas might come to mind:

- Active
- Vigorous
- Decisive
- Energetic
- A deed; the process of doing
- A series of movements or manipulations of the physical world

Thus, in this planning process, action is the embodiment of strategy; it is the doing, the movement, the physical energy that one lets loose in and operates on the world. And just as a boxer's punch needs to be timed right, aimed right, with the right amount of force—your actions will have many variables that can be tinkered with in order to keep them on target. A single strategy can lead to a number of actions, depending on the budget and the imagination of team members.

Look in on the Trojan Horse warriors again and evaluate how their strategy fared when they extended it into action.

After some elaborate storytelling and intervention on the Greeks' behalf from Poseidon, the God of the Sea (someone who may not be available to your company as a resource), the Trojans did bring the wooden horse into their city late in the day. It was left at the Temple of Athena, and they went off to bed.

Then, "In the middle of the night the door in the horse opened. One by one the chieftains let themselves down. They stole to the gates [of Troy] and threw them wide, and into the sleeping town marched the Greek army. What they had first to do could be carried out silently. Fires were started in buildings throughout the city. By the time the Trojans were awake, before they realized what had happened, while they were struggling into their armor, Troy was burning."[7]

A battle ensues. Greeks swoop down on the unprepared Trojans. Burning beams are thrown down on soldiers below, buildings and towers are toppled. The Trojans don armor from slain Greeks and, masquerading as Greek soldiers, attack bands of unsuspecting Greek warriors. In short, things take unexpected turns. Chaos happens.

Nonetheless, at the end of the battle, Troy lay in ruin, its heroes killed, its women widowed, and its children orphaned.

And what about Helen, the Greek beauty who had been kidnapped by the Trojans and was the original reason for the war? In the chaos, the Goddess of Love, Aphrodite, helps Helen escape safely and flee in a boat with Menelaus, her rightful Greek husband. But Aphrodite was not even at the strategic planning meeting! Without her help, would the Greek's plan have succeeded?

Was the strategy wrong? Was it just one more example of a good plan poorly executed? Or, after over a decade of fighting, had the Greeks simply lost sight of their original goal? What did they mean to accomplish?

Simply taking action is not enough; clearly, one must take the right action at the right time.

That is how the concepts of *strategy* and *action* take on a special meaning when they are brought together as *strategic action*. Together, they imply a joining of insight and energy.

Insight is drawn from our understanding of ourselves, our team, and our situation both inside the business and in the world outside. Insight is an essential component of strategy; it is the quality that gives strategy accuracy.

Energy is power; it is an essential component of action. Energy comes from the team itself and the leaders of the team and is based on commitment, motivation, and clarity of purpose. When these two strong forces of insight and energy are brought together in *strategic action*, the chances for success soar.

Strategic actions are effective because they have been crafted to take advantage of the challenges inherent in a situation, the resources at hand, and the skills, commitment, and innovation of the team members who will be executing them. That is why in the steps of the planning process, one does not separate strategy from action—they are combined into one powerful dynamic: action with insight (Figure 2.2).

Implicit in strategic action is the result or goal. A strategic action is an efficiently targeted activity undertaken to accomplish a specific objective. It is a way to optimize cooperative efforts to produce selected business results. By our definition of this concept, it is impossible to execute a strategic action without understanding clearly what its goal is.

The Greeks devised a creative and innovative strategy, and they followed it with vigorous action; but they were not taking *strategic action*. If they had been, the first warrior out of the horse would have opened the gates for their army; and the second would have already been on his way to Helen, enacting a plan for her safe return to Greece. Intervention by the goddess Aphrodite would not have been needed.

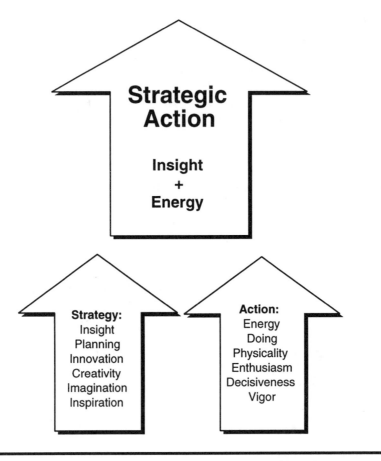

Figure 2.2 Strategic Action: The Strength of Joining Insight with Energy

These ideas will be readdressed as the book unfolds because they lay the groundwork for the success of your planning efforts.

NOW! A BRIEF INTERPRETATION OF TIME

But there is a another element that is critical to the process outlined in this text and that is the "NOW!" of the title.

"NOW" is this moment, then this one, then this one. Our "NOW" floats forward through time and provides us the most potent and influential place for making changes that have effect in the real world. It is this "NOW" that will allow you to benefit from the tremendous energy generated by your planning process. "NOW"—the continually unfolding present—is where the strategic action will take place.

And it is also this "NOW" that acts as a counter-balance to the theoretical, grey-matter world of strategy. Most projects have a deadline—the new product must be ready by March 1; the marketing campaign must take place before the holidays; the action plan must be in place before the budget cycle is completed. This is the pressure that "NOW" provides and also its power. Event-driven deadlines are the most effective.

The clock is always ticking and, in today's business world, the pace and pressure of "NOW" is almost tangible. And that is not always a bad thing; at some point, your team must stop planning and start doing. So "NOW" provides a fulcrum leveraging and balancing the brilliant strategists and the people who want to get busy doing.

But this "NOW" is not a floating bubble exempt from the influences of the past or the future. It is not exempt from that grey-matter world of thinking and planning. On the contrary, "NOW" is where the past, present, and future come together.

Everyone knows how one can be influenced by past actions, especially if one works with a team of people. Remember how Fred fouled up that important contract with one of your big suppliers last year? Now, Fred has been assigned to your planning project. How do you feel about it? How does Fred feel about it? Did he deny that he did anything wrong, and are others on the team angry with him because no one got the team bonus? Or did Fred lay it on the line and apologize to everyone for fouling up? Even if Fred did not admit his mistake in the past, he can still do it…now. But whether Fred 'fesses up or not, you will still be stuck with him on your team. And, who knows, it could be Fred's strategic idea that steals the show.

The reality is that the moment of "NOW" is not isolated from what has come before, any more than it is immune from the influences of what will come after it. In fact, one aspect of the entire process, whether one discusses the goal–strategy–action–goal cycle or the relationship of past, present and future, is that the planning process is integrated.

The aspects of thought and action are connected; you will not stop thinking and planning strategically when you and your team go on to devise your action steps. Even as your team is discussing challenges, your mind will be racing ahead to actions and wondering, "How are we going to *do* that?"

It is also likely in this fast-paced business environment, that your goal could need to be shifted after you have already begun executing your strategic actions. Conditions in your marketplace could radically change. You could lose a key member of your team. Throughout, you must keep all the links from one step to the others active at all times and the most effective place for keeping those links alive is "NOW."

Language is an incredible invention and tool but, unfortunately, everyone is somewhat hampered by the necessity of describing a process that progresses linearly from steps A to B to C to D when, in fact, time does not work that way, except on the page. But one is stuck with the descriptive process, and it can work as long as everyone knows that time is elastic, and that the steps of the process, although labeled as if they were discrete from one another and were happening in a sequence of boxes, are extremely permeable.

What follows is another in the set of exercises that will help your team progress through the planning process.

EXERCISE 2: INFLUENCES FROM THE PAST

Throughout the course of this book, exercises are offered that will take you and your team through all the steps of the planning process. The exercises are designed to give you—the project leader—help in organizing your meeting. They will outline the materials needed, estimate meeting duration, list team objectives, and provide facilitation and discussion tips. They are meant to be guidelines for your process; so do not be alarmed when your team's progress does not work exactly within the framework described or the timeframe suggested.

This particular exercise, "Influences from the Past," focuses one introspectively; you may want to do it alone first. If you then think it is necessary or appropriate, have your whole planning team participate.

EXERCISE 2: INFLUENCES FROM THE PAST

Materials: paper and pen.

Duration: 40 minutes.

Objective: To list features or issues of the past that may need to be taken into account in the course of your planning project.

Take a few moments to write down some of the critical influences from the past that could affect the project you are undertaking.

Think about the emotional and personnel issues, the facilities, out-dated equipment, a weak team member, budget cuts, new boss—whatever comes to mind that must be reckoned with in your project.

Just make a list. Do not worry about spelling, or whether things make sense. Turn off the critic and just brainstorm. Take out a piece of paper and start your list. Write until nothing else pops into your mind.

Probably the list you have made is mostly negatives, since that is what is on your mind and what you are most worried about. But do not forget the positives as well: past successes or similar experiences that can be drawn upon; resources created in the past that can be used now; the strength of individual team members; good team rapport; support from a key leader in management; or "chits" that can be called in from favors you did for others elsewhere in the organization.

So, if your first list was mostly negative, make another list that includes some of the positive aspects from the past. For now, do not make any judgments about what you should or should not include—simply write down your thoughts in whatever order they come to you.

Keep this list; it will be used as the basis for another exercise in Chapter 8.

Just as the past influences the present, the future plays a role as well. In fact, Russel Ackoff says, "Planning is the design of a desired future."[8] The future is what you hope to shape by completing your planning project. But the future, already existing in minds and ideas and present trends, will also be shaping you and your project.

Part of the reason for starting your strategic planning process is that you want to make sure your business can keep up with or surpass your competitors. You do not want to be caught resting on the laurels of a product that was good last year but is now old news. Maybe you are anticipating problems with a system conversion. Or you know that Gloria has been sending out her résumé and you cannot afford to lose your best product manager in the middle of your pre-holiday sales campaign. All these potentialities from the future affect your present as well. In Chapter 3, there is an exercise that isolates these future concerns.

The point is that "NOW" is where these events or ideas or emotions come together (Figure 2.3). This is the same "NOW" that will be the main arena for your strategic actions. Thus, it is important to understand the undercurrents—past and future—of the "NOW" as you begin to effect change and create the results you want for you or your business.[9]

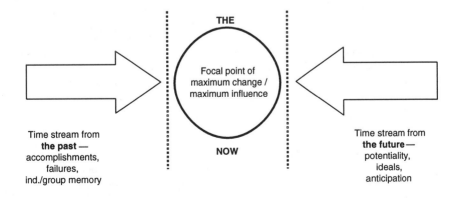

Figure 2.3   What is "NOW!"?

"NOW" has another connotation as well. You need to act quickly. The pace of change in the business world is accelerating. Ten years ago, did you predict what effect computers and the Internet would have on your business? Have you considered the ramifications of business-to-business commerce on the Internet? Or the revolution in consumer buying patterns and online shopping? What other invisible changes are brewing *now* that will have a direct effect on your corporation, your competitors, and/or your customers?

There is not a moment to lose. It is an exciting world of change and opportunity and your strategic action needs to be in place NOW!

The next chapter explores the characteristics and qualities of insight—a critical skill for a strategic planner.

NOTES

1. As quoted on the AMR Web site (www.advmfg.com), December 8, 1997. Malcolm McDonald is a Professor at the Cranfield University School of Management, England.

2. Information for this anecdote was taken from a *Wall Street Journal* story entitled, "Creating a Market in Missing Cups and Saucers," December 18, 1997, B1 and B2.

3. Information about the SEI transformation was derived from Fast Company, April/May 1998, "Total Teamwork," by Scott Kirsner, p. 130–135; and phone interviews with Mark Wilson, Corporate Team Member at SEI, in June 1998. In addition, SEI is one of several companies featured in *Driving Change: How the Best Companies Are Preparing for the 21st Century*, by Jerry Yoram Wind and Jeremy Main, The Free Press, New York, 1998.

4. I have a personal rule about change: three good-faith efforts. I use this whether I am trying to discuss an issue with a reluctant colleague, make suggestions in a performance review, or propose an idea that I believe in. After three good-faith efforts, if the door still seems closed, then I assume that pathway will not be functional and I reevaluate my options.

5. Ohmae, Kenichi, *The Mind of the Strategist: The Art of Japanese Business*, McGraw-Hill, New York, 1982, 37.

6. This story is rendered based on Edith Hamilton's telling of the Trojan War in *Mythology: Timeless Tales of Gods and Heroes*, The New American Library, New York, 1942. The primary chapters used are 13, "The Trojan War," and 14, "The Fall of Troy," p. 178–201.

7. *ibid.*, p. 199.

8. Ackoff, Russel L., *A Concept of Corporate Planning*, Wiley-Interscience, New York, 1971, 1.

9. My ideas about time are partially based on a discussion by Robert Sardello in his book *Love and the Soul, Creating a Future for Earth*, especially Chapter 4, "Time, Love, and the Soul," HarperCollins, New York, 1995.

3

BECOMING A STRATEGIST

One thing I learned: everything I needed to know was out there somewhere. The world glows with information.

Andrew Miller[1]

FROZEN STRATEGY VS. FLUID INSIGHT

In 1935, Leo Burnett, a former copywriter, journalist, and teacher, launched his own ad agency with eight associates in Chicago.[2] His approach was to build long-term client relationships, manage all aspects of a client's marketing and PR campaigns, and charge a hefty 15% commission. In the process, the agency created all-American icons that crystallized the Mom-Pop-and-apple-pie spirit of the times: the Jolly Green Giant, Tony the Tiger, the Marlboro man, and the Pillsbury doughboy. Sounds like a strategy for success.

And it was...for awhile. By 1975, the Leo Burnett Company celebrated its 40th anniversary and had become the world's fourth largest ad agency with $635 million in billings. The tag-line for one of its famous cigarette ads summed it up, "You've come a long way, baby." But long enough to rest on your laurels and have a smoke? Never.

Although the formula for success took a little longer than 40 years to collapse, of course it did. "Change is the only constant. Hanging on is the only sin."[3]

By the late 1980s and mid-1990s, it was more than apparent that the jig was up. Major, long-standing customers were going elsewhere. The creative and economic environment had changed dramatically since the company's inception. Clients struggling with the downsizing and belt-tightening of the 1970s and 1980s wanted a fee-for-service billing; they were looking for the best prices from specialty players rather than paying large umbrella retainers to a we-do-it-all shop.

With the computer and Internet boom transforming business culture in the 1990s and the infusion of younger executives, ad styles changed from Leave-It-to-Beaver to Beavis-and-Butthead. Graphics tended to be hip, slick, and irreverent. Sizzle and dazzle were in; safe and comfortable were passé.

By 1998, the Leo Burnett Company had fallen to 10th place in the world. Now it has begun to make radical changes—not only in how it approaches clients and how its business is structured, but on the creative side as well.

Looking back, of course, it is easy to say that the company lost sight of its customers' needs, lost touch with the culture, and failed to see changes that were more than apparent to newcomers in the field.

But hindsight is inferior to insight.

Insight is the ability to see the true nature of a situation, event, or person by perceiving what is essential or inherent there before an outcome makes it apparent. When one uses insight, one is thinking strategically.

This chapter is titled "Becoming a Strategist" because thinking strategically is not a static state. J.W. Marriott is attributed with saying, "Success is never final," and the Leo Burnett story could not be a more apt example of that. But do not misunderstand; Burnett was a brilliant strategist. He created a business empire that anticipated all the major trends in advertising in the mid-twentieth century and his company set the standard in the industry for decades. But every strategy must be tailored to each new business situation and, in this constantly changing world, every situation is different.

The Leo Burnett Company codified an approach and built a marketing organization that failed to allow the kind of strategic thinking that initially formed it to continue to inform its decisions. The need to think strategically never ends because business practices and the marketplace are always changing. At the most basic level, becoming a strategist simply means paying attention to as much as you can as much of the time as you can.

Strategic thinking is not acquired by magic. It is not something possessed by a select few or an elite group of seers. Insight is a talent and an ability that can be nurtured in all of us. Everyone has the capacity to think about actions and events of the past and to hypothesize about possible future outcomes. How well one does this is a matter of practicing and enhancing certain qualities in oneself.

The three aspects most critical to cultivating strategic thinking are:

■ The capacity to take in a range of information and the ability to understand the information accurately

- The pursuit of clarity, honesty, and integrity
- The courage to question, explore, and create

Each of these aspects for enhancing insight will be discussed separately.

INFORMATION HARVESTING: TREND-SPOTTING WITHOUT THE "POPCORN"

Anticipating the future is an inexact art. But one of the keys to formulating effective and actionable strategies is being able to anticipate what might happen in the future. The influences the future can have on the present have been discussed. How can you as accurately as possible understand what those influences might be? Where do you look for information that will help you?

Many people and institutions are known for their ability to spot trends, with greater or lesser accuracy. In fact, lots of businesses pay huge sums to professionals whose job it is to understand what is happening in the world, the country where they are doing business, and their own industries.

Faith Popcorn, John Naisbitt, Jennifer James, and others have staffs of people whose job it is to nose out trends and write about them.[4] There are firms of analysts who are paid to keep their clients informed about the latest business and technological trends in their industries. There are also well-financed and famous R&D think-tanks, like Palo Alto Research Center (PARC) or the Silicon Valley Incubator, where influential new products and services are being created based on ideas about what the future might hold.

But there is no reason to think that you cannot cultivate the same insight and skills that others have in seeing into the future, or one might more accurately say, seeing the future in the present. In fact, you may be better situated than any professional trend-spotter to understand the influences and tendencies that are present in your company or community or among your peers and colleagues.

Anyone who lived through the 1980s in corporate America witnessed a trend that started small and grew to enormous proportions and has finally been reversed or resolved, as all trends or tendencies generally are. Budget cuts, staff reduction, downsizing, and restructuring were everywhere: in the newspapers, in discussions at the copy machine, in books, and on talk shows. One did not need to track the Federal Reserve Bank's M1 or M2 figures to know that an economic downturn was happening. But the trick is to see a trend before it develops and, while it is developing, to understand its momentum enough to realize when the apex has been reached and another pattern is forming.

Some choose to distinguish between trends and fads, but no matter what one calls them, they are both tendencies of large groups of people toward certain types of behaviors. The difference is only a matter of degree or duration. In many cases, the trends and traditions in the business world are influenced by fads that emerge from mainstream culture.

Do not forget that the change produced by trends or fads is made by individuals—it is not technology or the economy that is sneaking up on us. Individuals are discovering, exploring, drawn to, or repelled by certain behaviors, activities, objects, or disciplines. *We*, all of us individually, are responsible for change, not some machine that we are merely watching.

In the 1960s, Beatle haircuts were a fad that stayed and set the tone for a decade of revolution in social behavior. Nonviolent protest, the women's movement, a relaxation of the dress code, a loosening of sexual mores were all significant events that spun out of the student fads of the 1960s and established the Baby Boomers as a group bonded to one another by their unique code of ethics, values, and preferences.

These trends in behavior set the stage for a loosening of the traditional ways of doing business that had been the standard since the turn of the twentieth century. The 'good-old-boy' network, fierce competition and territorial battles, business secrets (many times even within an organization), and centralized management and control are now being replaced by collaboration, business partnering, decentralization of information, and business unit autonomy and interdependence.

Other changes, like the revolution in information technology precipitated by the inventions of the telephone and personal computer, for example, have a longer life cycle and build more slowly. These changes are more fundamental and affect the structure of our lives because the tools used for living are being revolutionized. The long-term effects of the current technological revolution are decades or even centuries from being fully felt or understood. They have been the catalyst for much of the innovation in the business world in the last 10 to 15 years and will continue to drive change well into the middle of the next century.

In an article by Harvard Business School professors C.K. Prahalad and Gary Hamel, the authors talk about the need for businesses not only to consider the improvement and enhancement of current consumer products, but also the actual creation of products "which customers need but have not yet imagined."[5] Now that is strategic thinking.

But the point here is that all of us are potential receivers of trend information. Potentialities and tendencies are apparent in all aspects of our lives. They are apparent in our local newspapers. They can be heard in conversations on poker night. Our children will come home talking about them or wearing them. We are the change agents, and all of us are experts about our worlds. The key is to be open to what is happening around us and to understand the significance of the clues we are being given.

EXERCISE 3: INFLUENCES FROM THE FUTURE

In order to focus on the current trends in your environment that might affect your planning project, take a few moments to reflect on what can be called *influences from the future.*

Remember, predicting the future is an inexact art. Do not be afraid to use your intuition or insight.

EXERCISE 3: INFLUENCES FROM THE FUTURE

Materials: pen and paper.

Duration: 20–30 minutes.

Objective: to outline the influences from the future that might affect your project.

Take a moment to jot down ideas, intuitions, or clues you may have about influences from the future that could have an effect on the planning project you are beginning.

Think about not only the things you see inside your company—potential personnel problems, merger rumors, shifts in product development direction—but also things in the world around you that might come into play.

Is the major economic base of your region shifting?

Is the ethnic make-up of your sales region changing?

Try to suspend your judgment for now about whether an idea seems crazy or unfounded. Just allow yourself to make a note of whatever comes into your head.

Hang on to this list. We will return to these ideas in Chapter 8.

WHAT IS ON YOUR READING LIST?

One way of staying tuned in to potential future influences is to take in a wide range of information from a variety of sources. Inventor Stanley Mason admits that he has a stack of magazines three feet high on his work desk.[6]

I must admit I, too, am much more comfortable than many people with the written word as a source of information for a variety of reasons: its authenticity and accuracy can be checked; generally, it has a byline and a responsible author; there are clearer legal parameters for written documents (the verdict is still out on how to deal with "virtual" information); and faulty reasoning is easier to pick out in writing that must conform to certain conventions of grammar and syntax.

So, perhaps the first thing to look at in considering how you take in information is your reading list. It is important of course to keep current in your field or specific industry by subscribing to trade magazines or publications. Even if you do not have time to read all of them thoroughly, scan the headlines or the table of contents. Glance at the contributors' notes. Look at the advertising. What topics are being covered? Who is writing about them? Who is paying to get your attention? What selling points are the ads focusing on? What do the ads look like graphically and how do they make you feel?

You are probably also reading one of the regional or business dailies—the *Wall Street Journal, Business Investor, New York* or *Los Angeles Times*. And maybe a local paper as well. (If you do not want to spend the money for a lot of subscriptions, but have a computer and modem, check out the online resources. Many of these publications have very robust Web sites).[7]

Whatever your field, no one in business these days can afford to ignore high tech. You may already be reading one of the subscription-only distributed trade magazines: *InformationWeek, InternetWeek* (formerly *CommunicationsWeek*), or *Inter@ctive*. You may also want to consider one of the technology magazines that takes a broader view of the subject: *Upside* (the readable version of *Wired*) is a good one. The *San Jose Mercury News* gives excellent coverage of Silicon Valley happenings. The *San Francisco Business Times* is a useful weekly for the same reasons. Another good Bay Area, local high-tech magazine is *Computer Currents*; it provides a heard-on-the-street approach to the computer industry.

For innovative business ideas with a high-tech slant, try *Fast Company, Red Herring*, or *Entrepreneur*. Publications about the Web are breeding like mosquitoes. One of my favorites is *Boardwatch*; it has a range of topics from Internet to bandwidth to telecom and has been around long enough to stay around.

One way to be efficient about your time is to make sure that you have one or two selections on your reading list that review or excerpt from other media. One of the best is the *New York Times Book Review* (part of the Sunday edition of the *New York Times*). Always full of a range of subjects, these well-written synopses of recently published books cover a lot of ground. They show not only the range of book topics being published—whether sociological, scientific or political—but the themes and ideas that seem to be popular in all types of writing, including both fiction, non-fiction, poetry, children's books, mysteries, and large-format art books. The Best Sellers' Lists give a snapshot of what a lot of people are reading in hardcover and paper, fiction and non-fiction. This one publication in itself has a wealth of trend information.

Or consider a compiler like *Utne Reader*, which reprints selections from the best of many small press publications; the people at *Utne* do some of your trend-spotting work for you by grouping topics for each issue under a specific theme.

A newsletter like *BottomLine Personal* is also a good choice; it has a wide range of information on all topics applicable to a modern reader: investment and financial advice; how to deal more effectively with your spouse, children, or pets; medical information; good deals on products, services, or vacation spots; free literature to send for; garden tips; and safety advice, among other topics.

These are fairly obvious choices; but it is always a good idea to stray outside of one's familiar reading territory from time to time. Pick up a large-format tabloid like *Rolling Stone* or *Interview*. See what is happening on the pages of *Vanity Fair*. Get a copy of the lush (and pricey) *Communication Arts* or *Architectural Digest*. Look at one of the fine small literary magazines like *Granta* or *Zyzzyva*. Grab a local art and culture magazine from your area or even a 'zine—a small-circulation, home-made comic or underground magazine. (These can usually be found at a counter-culture store or a good local bookstore, not one of the national chains.) Look not only at the content but also the graphics and layout. You never know where your ideas and inspirational insight for trend-spotting will come from.

And, in my opinion, the *New Yorker* is simply one of the most consistently excellent and broad-ranging English-language magazines around. A must-read.

Since ours is a media culture, a modern strategist cannot afford to overlook other forms of information exchange. Listen to the radio. Take yourself to an art gallery or the local museum. See some live theatre or dance theatre. Find the local venue for live music and listen to what is happening. What is being displayed or talked about? What is popular or stylish? What styles are merging or re-emerging?

And do not make the mistake of ignoring what is going on internationally. If you watch only the U.S. news, you will hear about natural disasters or assassinations, but not much else.

Pick a country or two that interests you and follow its news in as much detail as you can find (read the *Economist* for great international coverage)—not just the crises, but the daily politics. Choose a few political leaders that interest you and become familiar with their policies and constituencies. Learn a bit about the culture and the language. Pick up a newspaper or magazine in another language when you are traveling. Even if you cannot read every word, you can probably scan the headlines, look at the graphical style and format of information presentation, and "read" the photos.

Of course, traveling is a sure-fire way of seeing firsthand what is going on in the world; and more than that, it often puts one's own culture in stark relief so that the mundane and taken-for-granted suddenly become visible.

The point is to broaden your vision—that is part of being an informed citizen of the world and a good strategic planner. Do not bury your head in the corporate sand pile. There is a big world out there and your business is in it.

CHI: CLARITY, HONESTY, INTEGRITY

The second aspect of exercising your insight is what I call CHI (pronounced "chee"). In martial arts, CHI is a special kind of inner energy that is a combination of physical, mental, and emotional readiness for action.

Traditionally, CHI has many derivations and exists in several Asian cultures.[8] The Chinese believe that *Ch'i* or *Qi* was introduced as a concept by the philosopher Mencius, born 100 years after the death of Confucius in the fourth century B.C. Mencius taught that man is innately good and that one's nature can be enhanced or perverted by one's environment. His notion of *Ch'i* is air or breath, the universal energy that exists in all things.

The Japanese *Ki*, the Vietnamese *Khi*, the Indonesian *Ihru*, and the Chinese *Ch'i* all refer to the vital energy of life, the creative energy, the divine breath that appears in action, attention, and concentrated mental force. These are all ideas related to insight and the cultivation of strategic thinking.

I want to acknowledge these traditional meanings of CHI at the same time I propose a Western application of this ancient Asian concept. My use of CHI is a symbolic acronym for **c**larity, **h**onesty, and **i**ntegrity.

But how, you might ask, are these qualities relevant for strategic action? How is being honest going to get me anywhere in the corporate world? I discuss these values as a way of offering some, perhaps old-fashioned, ideas about the kind of self that makes for an effective strategist.

Clarity translates to accuracy of judgment and the ability to understand information without the shadowing filter of exaggerated emotion, personal ego, or misapprehension. No matter how much information you gather or how wide ranging your gathering, the information will be useless if you are unable to understand it clearly. And I believe that you cannot understand information clearly unless you have a clear understanding of yourself and your place in the world.

In an extreme example, one can imagine a paranoid deducing from every detail of his day that there is a plot on his life. The neighbor next door closing his curtains during the day (because he is a night worker and sleeps during the day) is storing up weapons and plotting revenge. A new security guard posted at the corner convenience store (because of a recent string of robberies) is there to train the neighbors in assault techniques. The smell of gas in the kitchen (because one of the pilot lights is off) is a conspiratorial attempt by the utility companies to catch our victim off-guard.

Clearly, these same bits of information might result in a very different set of conclusions when perceived by a person who is not convinced that unknown persons in the world are out to do him or her harm.

Granted, this example is an exaggeration, but it is used simply to illustrate that if information is not read accurately, it is useless; in fact, it can be more than useless—it can be dangerous. Misapprehended information can block your view of what is really going on. That can have such diverse and dire effects as not seeing your partner's divorce request coming, ending up in the wrong place at the wrong time, or not realizing until too late that your business needs more capital to survive.

In the same way that one must have an accurate view of one's business internally—operations, personnel, finance—and externally—the local and global marketplace, competitors, changes in the industry—one must have an accurate view of one's self. So, in citing clarity here, I am also referring to self-understanding and accuracy in perceiving the self, both in the world and as an operant in your business.

The next two qualities of CHI—honesty and integrity—are moral values that inform the way one acts or does business. Honesty is something everyone strives for, and it is certainly given much lip service as a highly valued personal characteristic. It is something we teach our children from a very early age.

In the context of becoming a strategist, honesty is striving for the truth. That may sound highfalutin, but it really is not. It simply means being a straight talker and calling things as they are. Striving for truth also has to do with self-honesty. Knowing one's limits, one's biases, and one's faults is a difficult and necessary part of being a good manager of others and an insightful strategist.

But how does one act on the value of honesty in the business world? Is it really practical? That same child we have impressed with telling the truth at all times learns very early that there are instances when other considerations must be taken into account. One does not say that Grandma Gertrude is wearing a perfectly ridiculous hat. In the same way, anyone in the corporate world knows that there are mitigating circumstances that must be balanced in an effort to act honestly and speak the truth.

Barbara Ley Toffler, Partner and Director of Ethics and Responsible Business Practices for Arthur Andersen in New York, and her husband, social ethics scholar Charles W. Powers,[9] refer to this weighing of ethical options as exercising one's "moral imagination." We often hold more than one value at a time, and sometimes, when they conflict, we cannot exercise each of those values as fully as we might like.

Toffler gives the example of a manager who has been told some disturbing information by an informant who wants his identity protected. This situation leads the manager to confront another of his employees by waving an unrelated memo and saying, "I have an anonymous letter here stating…What can you tell me about this?"

The ethical dilemma is whether lying about having the letter justifies establishing the truth or falsity about activities, possibly perpetrated by an employee, that are illegal and unsafe. In this instance, the manager has many values to balance: his wish to protect the safety and honor the confidentiality of his informant; his own desire to be honest; his need to discover the truth in order to minimize the risk to his company and his employees; his obligation to stop an illegal activity, if one is occurring; and his desire to deal fairly with the alleged perpetrator of the activity.

Did the manager stretch the truth, tell a "little white lie," bend the means to justify the end? What would you have done? How creative would you have been in exercising your 'moral imagination' to come up with a solution?

In another less dramatic example, there are times when information critical to a product's success is impossible to share with others; but this is not dishonesty, this is protecting the interests of one's company. On the other hand, one would have to say that the cigarette executives lined up at the Senate hearings, swearing, one after another, "Cigarette smoking is not addictive," were committing acts of gross dishonesty.

It is important that you operate from your best and strongest self. When you are required to compromise by speaking less than the truth day after day or by taking actions that you feel are unethical, a little bit of the best of you dies. If one has to make a judgment call about when to speak the truth and when not to, something is already being compromised.

In a recent article on honesty in *Fast Company*, Chuck House, Executive Vice President of Dialogic Corporation, Parsippany, New Jersey, who also spent three decades at Hewlett-Packard, developed this motto: "Come to work each day willing to be fired."[10] One might say that honesty is the principle and having the integrity to speak the truth is putting that principle into action.

The truth does not have to be brutal, however. One of the challenges in speaking the truth is finding gentle ways to discuss difficult truths. Honesty requires subtlety and, often, kindness. As Michael Wheeler, Professor of Management at Harvard Business School, says, "Blunt questions can force people into corners where they feel compelled to shade things—even to lie. Instead of saying, 'Are you in favor of this project?' you should ask, 'How can we improve it?'"[11]

This implies that there are ways to encourage truth in others as well as in yourself. And it points out how, on an individual level, the values of a corporate culture are made real.

In the same way that an individual can be deceived about strengths or possess an attitude that blocks understanding, so can an organization. Corporate culture is a system of shared beliefs and values held by the members of a given company. It is obvious, but worth saying, that corporate culture is made up of individuals and their values. On the other hand, it is difficult to be honest in a culture that does not reward or encourage honesty and this corporate dishonesty can prevent the accurate assimilation of critical information.

Apple Computer had the right idea at the right time but made several critical errors in judgment that allowed the company to languish on the brink of disaster. In a sense, Apple did not realize soon enough what their real product was. They did not understand that their corporate strength was not their computer product but their creativity and innovation in the way they conceived of the computing machine. They missed the opportunity—that Microsoft grabbed—of understanding the importance of the operating system.

There are many who believe that Apple's corporate arrogance—or one might say their dishonesty in evaluating their strengths and weaknesses—is what prevented them from understanding more clearly how to position the company for sustainable growth. It seems to me to be the same kind of arrogance that Microsoft is now exhibiting in its attempt to coerce buyers of its operating system into using its browser. At some basic level, it does not matter if the product is better or not; people want to be treated with respect; they want to make their own choices. It is the same arrogance or dishonesty or failure to search after truth that caused the Leo Burnett Company to temporarily stumble.

Integrity is sound judgment coupled with honesty. If honesty and integrity are truly values in a company's corporate culture—and not just words in a mission statement—individuals can more easily speak out about attitudes or projects that are destructive to a company's strategic direction. If you and your business colleagues have strong CHI, your corporate culture and your corporation will be stronger.

In fact, Toffler says, "Ethics is an energizer. I've seen the positive bottom-line effects of ethical behavior. People become more effective decision-makers and implementors. They work together differently."[12]

So perhaps this discussion of values might seem out of place in a strategic planning text, but as the business world reacts more quickly to changes in information technology and innovation, it is critical to reinforce the foundational values of your corporation. It is these values that influence the day-to-day workings of your organization, and it is my contention that the values to which you and your company adhere provide an environment more or less conducive to clear strategic thinking.

To summarize, there are several general precepts I have found in my own business dealings that can both elicit and nurture one's CHI. They include:

- Understand your limitations, professionally and emotionally: if you are forthright about your limitations, others will be too and the right mix of abilities will be easier to assess and bring together in your team or work group.
- Be as clear and honest as you can about bad news or obstacles to success—denials make things worse because the real news always breaks out in the end and then you look like a cad. Often, the 'bad news' sparks a new direction or idea that offers a surprising opportunity.
- Be sincere and generous with praise.
- Conduct yourself honestly and with personal integrity. Although there are not always immediate or evident rewards for right action, it is always best; you will generally know what it is and, if you have the courage to do it, you will sleep better.
- Hope for the best but assume the worst—that usually provides for a path somewhere down the middle; reality seldom operates very long at the extremes.

EXERCISE 4: NURTURING YOUR CHI

You may have your own ideas about how to keep your personal CHI healthy. Or maybe you have not thought much about what it is that makes you physically, mentally, and emotionally ready to take action. But do not take your position or your mental and emotional well-being for granted. Many people look to you and count on you as a leader in your organization. So take a moment to acknowledge that to yourself. Then consider these questions. What strengthens you as a leader in your organization? What values do you cherish or strive for?

Take a moment to explore your own values in the following exercise. This exercise is not for the team, but for you as team leader.

Just a word or two about these exercises. If you are anything like me, you will be tempted not to actually do the exercises. You may read through the exercise questions and feel that that qualifies as completing the exercise; or you will think, "Well, I get the main idea. I see what she's after; so why do I need to actually do it?"

But, in fact, a different process and experience happens when you actually sit with pen in hand (or computer keyboard) and write out answers to these exercise questions. Writing is not a matter of becoming a scribe for your own mental processes; that is, there are no preformed answers waiting in your brain to be captured on paper. The process of writing is a physical and mental collaboration. It is an exploration, using language as a tool, to discover something about yourself that has not been revealed before. And simply reading a question may start you thinking, but actually doing the exercise is what manifests the benefit of the thinking process that takes place.

Thus, you will get more benefit from this book if you and your team actually do the exercises included. End of pitch.

EXERCISE 4: NURTURING YOUR CHI

Materials: pen and paper.

Duration: 30–45 minutes.

Objective: to allow you to focus on the values that nurture your chi; and to explore your preparation for leading your planning project.

Read through each of the question groupings below. When you finish reading, take out a paper and pen, or your laptop, and write your answer for each set of questions.

1. CHI is an acronym for clarity, honesty, and integrity. Are there other values that are important to you personally or professionally? How do you nurture these values in yourself?
2. Can you think of an action you took at work that reinforced one of the values you listed? What was it? Did it involve other people, or was it something that happened inside of you? Describe the event and the feeling you have about it.
3. Are there times when you feel you need to compromise your own integrity in your workplace? Describe one such event in the recent past. Why did it make you uncomfortable? Can you imagine a way it could have been handled differently?
4. Now take a few moments to mull over your strengths as a strategic thinker and leader, particularly as they relate to the qualities discussed here.

Where do you get your ideas or inspirations? And, in the spirit of your CHI, are there ways in which you could improve your capacity for insight and your abilities to lead? What are they? What do you feel confident about and what would you want to improve?

5. Are you ready to take action on this planning project? If not, what would you do to be better prepared?

THE COURAGE TO CREATE

Anyone who has a two- or three-year-old child or grandchild will suddenly realize how much they do not know about the world.

- Where does light come from?
- Why is your hair red?
- Do ants have pets?

These and thousands of other questions that we have never thought to ask take on topmost importance and, even after an answer is given, are generally followed by "why?". Although this string of questions drives most adults crazy, in fact, the mind of a child is the mind of a strategist. One of the roots of insight is the ability to question, explore, and create. A true strategist does not shy away from a string of 'whys' and must be willing to ask "what if..."

If someone on your team says, "But we can't increase our sales volume," you must ask, "Why?" And when the answer is, "Because we can't lower our product price," you must be willing to ask, "Why?" And when the answer is, "Because our suppliers' contracts for X fix our price at $Y," you must be willing to ask, "Can our product be made differently?" or "Can we partner with a different supplier?" or "Can we make X obsolete?" or whatever is necessary to fully explore the possibilities.

Sometimes, a string of whys can serve to unearth the real issues or concerns. If you are getting resistance in a planning session, the 'string of whys' technique can sometimes be a catalyst for provoking a new way of thinking. Sometimes it provokes anger or indignation—but these feelings, too, if they are followed, can ultimately result in some innovative answers.

Creating an idea takes courage. It is so much simpler to remain silent and so much more boring! Be bold; consider options. The strategic thinking process requires self-confidence, determination, and a child-like curiosity.[13]

How else could Orville Wright have been convinced that something heavier than air could fly? Or Marie Curie discover and study an invisible substance? Or Leonardo da Vinci make drawings for a vessel that would maneuver underwater?

How else except by asking, exploring, and creating. Further discussion about the qualities that encourage creativity is given in Chapter 8: Strategic Action.

So, now that you have been reminded of your abilities to exercise and nurture strategic thinking, get ready to begin your project. The next chapter outlines the four-step planning process.

NOTES

1. Miller, Andrew, Bookend, *New York Times Book Review,* Oct. 12, 1997, 39.
2. The information for the Leo Burnett story was taken primarily from an article in the *New York Times,* "Not So Jolly Now, a Giant Agency Retools," Business section, p. 1 and 7, March 22, 1998; and other online sources.
3. Denise McCluggage, race car driver, quoted in *Women's Sports,* June 1977, p. 18.
4. For more information on Jennifer James, the urban cultural anthropologist, call Enterprise Media, 1-800-423-6021, or email her at jjanthro@msn.com. Her books include *Thinking in the Future Tense,* Touchstone Books, Simon & Schuster, New York, 1997, and *Twenty Vision Steps to Wisdom,* Newmarket Press, New York, 1997. Faith Popcorn is the best-selling author of the *Popcorn Report* and the Chair of BrainReserve Inc., a trends consulting company that she founded in 1974. *Clicking,* HarperBusiness, New York, 1997, is her most recent book.
   John Naisbitt is co-author with Patricia Aburdene of *Megatrends Asia,* Simon & Schuster, New York, 1996; *Megatrends 2000,* William and Morrow Company, Inc., New York, 1990; and *Re-Inventing the Corporation,* Warner Books, New York, 1985, and others. His Web site is only an advertisement for his books, but can be found at www.naisbitt.com.
5. From an article by Harvard Business School professors C.K. Prahalad and Gary Hamel, "The Core Competence of the Corporation," *Harvard Business Review,* May/June 1990.
6. Stanley Mason, CEO of research and development firm SIMCO, in Weston, CT, and Professor of Entrepreneurship at Sacred Heart University in Fairfield, CT, is the inventor of the disposable diaper, the easy-open bandage package, trash-compacting systems, the snack bar, microwave cookware, and many other household items.
   Mr. Mason's book, *Inventing Small Products, for Big Profits Quickly,* Crisp Management Library, Menlo Park, CA, 1997, is not only about invention but about the process of creativity. One of my favorite Mason axioms is "Aim for excellence, not perfection...Excellence takes minutes or hours or days. Perfection may delay you forever." (p. 79).
7. Listing of homepage URLs (universal resource locator = address) for the magazines mentioned, if they exist:
   BottomLine Personal: unfortunately, they do not post the text from the newsletter; you must buy a subscription; www.boardwatch.com
   Communication Arts: www.commarts.com
   Computer Currents: www.currents.net
   Economist: www.economist.com

Entrepreneur: www.entrepreneurmag.com
Fast Company: www.fastcompany.com
Granta: www.granta.com
InformationWeek: www.informationweek.com or pubs.cmpnet.com/iw/683
Inter@ctive: www.interactive-week.com/~intweek
InternetWeek: http://techweb.cmp.com/internetwk/
NYTimes Book Review: www.nytimes.com
Red Herring: www.redherring.com
Rolling Stone: www.rollingstone.com
San Francisco Business Times: www.amcity.com/sanfrancisco
San Jose Mercury News: www.mercurynews.com
Upside: www.upside.com
Utne Reader: www.utne.com
Vanity Fair: www.vf.com
Wall Street Journal: www.wsj.com
Wired: www.wired.com
Zyzzyva: www.webdelsol.com/ZYZZYVA/

8. Information on the traditional Asian use of the word "ch'i" was taken from a variety of sources: *The Original Martial Arts Encyclopedia*, by John Cocoran, Emil Farkas with Stuart Sobel, Pro Action Publishing, Los Angeles, 1993; *A Dictionary of Martial Arts*, by Louis Frederic, Charles E. Tuttle Company, Inc., Boston, 1991; and *Martial Arts—A Complete Illustrated History*, by Michael Finn, The Overlook Press, Woodstock, 1988.

9. Barbara Ley Toffler Ph.D., Partner and Director of Ethics and Responsible Business Practices for Arthur Andersen L.L.P. in New York, can be reached at 212-708-4903 (voice) or 212-445-9607 (fax), or by post at Arthur Andersen, L.L.P., 1345 Avenue of the Americas, New York, New York 10105. Her book of interviews with 33 corporate managers talking about ethics is entitled *Managers Talk Ethics: Making Tough Choices in a Competitive Business World* and is available by writing to Arthur Andersen, New York. The ethical dilemma used in our text is taken from this book.

   Charles W. Powers Ph.D. is Executive Director of CRESP (Consortium for Risk Evaluation with Stakeholder Participation) and works with EOHSI (Environmental and Occupational Health and Safety Institute) for the New Jersey School of Medicine and Dentistry, Piscataway, NJ. He co-authored, with John G. Simon and Jon P. Gunneman, *The Ethical Investor: Universities and Corporate Responsibility*, Yale University Press, New Haven, 1972 (currently out-of-print); and wrote *Ethics in the Education of Business Managers*, John Wiley & Sons, New York, 1991 (currently out-of-print).

10. Taken from "The Truth Is, The Truth Hurts," in *Fast Company*, April:May 1998, 94. Chuck House is Executive Vice President of Dialogic Corporation in Parsippany, NJ. Mr. House can be reached at c.house@dialogic.com.

11. ibid., 96, Michael Wheeler is a Professor at Harvard Business School, Boston, MA. He can be reached at mwheeler@hbs.harvard.edu.

12. Ms. Toffler's comments quoted in the text were taken from several phone interviews with her in early June 1998.

13. Stanley Mason's book, *Inventing Small Products, for Big Profits Quickly*, includes a wonderful section of axioms and tips for sparking creativity.

4

STRATEGIC ACTION
PLANNING: FOUR STEPS

In truth, most good plans come from the gut, depending on analysis only to uncover basic flaws and underlying external threats and to document what instinct already knows.

C. Davis Fogg[1]

PLANNING METHODOLOGY

If you peruse the 400 citations from *Books In Print* that have a strategic planning focus or you sign on to *amazon.com* and scan the over 1500 hits attributed to "strategic," you might well ask yourself, "Why do we need another book on strategic planning?"

I think the answer is elucidated in the quotation above, and the irony is that in few other planning texts has the obvious been stated so clearly. We need another planning book because there are few texts that outline a tactical process that *is* based on common sense, good basic business communications skills, and intuition. This planning method is based on the notion that planning is most often a process of revealing or confirming what you already know about your business and its marketplace environment.

The classic strategic planning process, born in the 1950s and 1960s, used specialized personnel in isolated departments. During the intervening decades, planning has evolved into a less cumbersome procedure. Because of the radical shifts in the business world brought on by technology and the demand for greater efficiency—both of which are transforming the structure of businesses—more organizations are requiring that all managers have basic planning skills and be able to execute collaborative planning

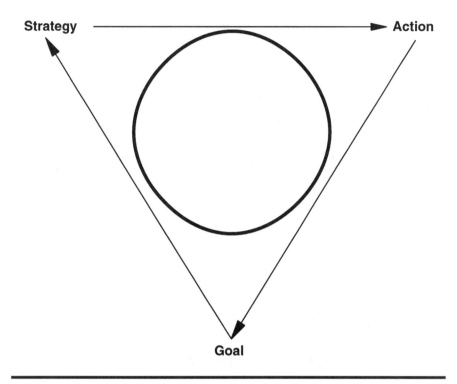

Strategy ⟶ **Action**

Goal

Figure 4.1 Basic Planning Cycle

initiatives. Planning is not isolated in one department anymore; nor is it the special province of one set of employees. Everyone in an organization has to have some knowledge of planning techniques. The team-based tactical process outlined here is assumed to be only part of a larger strategic effort that is taking place in your organization.

This chapter describes, in overview fashion, a four-step planning methodology so that you will have the whole process in mind when you begin facilitating with your team. It begins by clarifying a few terms and adding some detail to the basic planning cycle model that was discussed in Chapter 2: Beginning Concepts.

Recall that this model outlines the basic steps in any planning cycle: the first step is to formulate a preliminary goal, generally by conducting a situational analysis; the next is to devise some strategies based on relevant conditions or constraints; and the final step is to take action on those strategies in the hopes of reaching the stated goal (Figure 4.1). I have differentiated the final goal from the preliminary goal here because although you target your strategy based on your initial goal, often in the process of either devising or implementing strategic action, the goal

1. Challenge	2. (preliminary) Goal	3. Strategic action	4. Monitoring & measurement	(final) Goal
Why are we planning? What needs changing?	*What* do we hope to achieve?	*How* will we do it?	*What* will we track? *Who* will do it? *When* will it be done?	Did we make it? What needs changing?
Situational analysis	Objectives	Strategy	Control	Result/ Situational analysis

Figure 4.2   Strategic Action Planning Cycle

becomes refined or redefined such that the actual result—the final goal—is slightly different.

With the addition of some finer detail and commentary to the process, the major milestones might look something more like Figure 4.2.

The top row—numbers 1 through 4—are the basic planning steps of the process. Note that the final box indicates the result or final goal, but it can also be the beginning of the next planning cycle. So, although the steps are illustrated in a chart, they are actually comprised of stages that lead into one another and represent a continuous cycle of planned change.

The second row lists the questions that one might ask to complete the steps outlined above. The last row indicates how these steps might be referred to in other planning methods or what steps they correlate with in a more formal strategic planning process.

Remember that although these steps appear in neat boxes here, this is only to simplify one's understanding of the general territory of each task and the ideal completion sequence for the steps. In the real world, these steps overlap and blur into one another both backward and forward.

At any rate, these four steps become the skeletal model for the planning procedure. When I work with clients, I call them Cate Gable's Strategic Method or CGSM©.

The four steps are the following:

■ Challenge
■ Goal
■ Strategic Action
■ Monitoring and Measurements

Each one of these major steps, although labeled here with a single word, actually represents a complex set of subtasks that you and your team will accomplish before moving on to the next categorical step. Each of these four steps will be discussed in its own chapter or chapters, and each step will be broken down into more detail as we zoom in to scrutinize it at the task and subtask levels.

Before going on to describe each of these major steps, I want to give you the thinking behind the process used in this book. Several techniques have been developed to help you and your team through your planning project:

- Discussion of terms and concepts
- Exercises
- Facilitation tips
- Sample planning items
- Pointers for further study

Terms and Concepts

Every process needs a vocabulary. In a team-based planning project, language is the primary tool. For the specific purposes of this planning procedure, we must agree on what certain terms and processes will be called, or what we mean exactly when we use a certain word or phrase. You and your team will begin to develop your own language and way of communicating together as you go through the process of building your plan, but you will start within the framework of this four-step process and its language.

There is nothing technical about the language used in this book. As you will see, there is no academic discussion about the difference between an objective and a tactic. But there will be discussion of the language used to designate activities and an exploration of what heritage of meaning some words carry with them. This is simply a way of acknowledging that we must be clear with each other and of confirming that we have the same understanding of what we are doing. To provide you with a quick-reference guide to meanings, there is a glossary in the back of the book.

Exercises

The planning procedure provided in this book assumes that you will be working through your project with a team; these exercises outline the agendas and scope of tasks for team meetings. These exercises are framed to allow your team to accomplish the tasks required for the completion of each of the four major planning steps. You may find that you need additional meetings to supplement or complete the exercises included here; but the main steps in your group process can be accomplished by doing the provided group exercises.

The Appendices include a summary of the entire planning process and the names and page locations of all the exercises in their proper sequence.

Facilitation Tips

Embedded in the exercises from time to time are tips to help you as leader of this planning project and as meeting facilitator. These are handy techniques that I have acquired from others or created myself in the course of many meeting and workshop facilitations. They help with the administrative details of a meeting or problems that arise when you get a group of people together: How do you make sure small-group work ends at the same time? What do you do with team members who are always late? How do you handle comments that sidetrack the task at hand? etc.

Sample Planning Items

At the end of each chapter covering the four major planning steps, you will find sample items that might have resulted from a hypothetical team discussion following the group exercises included in that chapter. These sample planning items are developed over the whole four-step process. After each exercise, the same sample items are transformed and carried through to the next step.

For example, in the chapter on identifying challenges, there is a sample list that might have been developed as a result of that step. These sample challenges are sorted, then carried forward to become goals. Next, they are addressed in the strategic action step, and, finally, monitoring and measurement components are added.

These samples are intended more for you, the project leader, than for your group. They are included to give you examples of how the process might work. Your team will be generating its own list of challenges.

Pointers for Further Study

The chapter notes not only give the citations for quotations and information used in this book, but also include comments about materials that you might want to look at if you have time to study further. In addition, at the back of the book, there is an item called "Selected Readings." This section is divided by topic and includes comments and recommendations for one or two books that deal specifically with the topic indicated. Most strategic planning texts have such a lengthy bibliography that it is difficult to know just where to start if one does not have a lot of time to sort and read material.

THE CHALLENGE

The challenge is the WHY in your process, in that it is the reason you are planning at all—something needs to be changed and that becomes the catalyst for your action.

Another way of understanding exactly what is meant by your business challenges is to give them the more daunting label of problems or difficulties—or you might, as many companies do, call them opportunities. So, the objective of the challenge step in this planning process is to identify the difficulties/opportunities/challenges facing your business today.

This phase of the process is a grassroots assessment or situational analysis. During this initial discovery phase of planning, these are the kinds of questions that might spark your team's thinking.

- What do you need to change, overcome, or fix about your business? your department? your marketing effort? your product? your customer service? your manufacturing or fulfillment procedures? etc.
- Are you being forced to change your product direction because of international competition? What are the challenges facing your product in the international marketplace?
- Is the state of your communications infrastructure robust enough to support the demand for efficiency internally? externally?
- Do you need to increase your sales by developing new target markets or by using additional means of distribution?
- Are your sales growing, level, or decreasing? Do you know why? What about your competitors' sales?
- If your strategic direction or vision has been made clear to you from top management, what does your department need to improve or take charge of to support those corporate goals?
- What resources do you need to keep up in the marketplace? Do you have the resources you need to make necessary changes? How can you ensure that you have the proper resources?

My guess is that most of you will not need to conduct a long, detailed study about the challenges facing your department. Either you have a pretty good idea what they are, or corporate challenges have been discussed with you already by the upper management in your organization, and these corporate challenges are what you will be supporting as you devise strategic action in your planning sessions.

I have never been in a business meeting in which the 'issues or concerns' agenda item left everyone happily smiling with nothing to say. Much more likely is that most project teams have many more challenges than they can successfully handle within the limits of resources and time available. If that is the case for your team, then one of your jobs will be to decide which challenges are relevant or appropriate for your team to tackle or which ones will give the best corporate payback for the effort.

Another of your tasks will be to understand which challenges may be related to or the causes of other difficulties; or which challenges may mask other more profound problems that have been left unstated. The next chapter, called "Seeing the Challenge," begins the challenge step in the planning model: methods for exploring, sorting, and prioritizing the challenges that you and your team identify will be discussed.

This list of challenges that you generate from the exercises in the next chapter will be the starting point for your planning project; these challenges will carry through into all the following steps of the planning. So, in one way, you might say that this first step is the most important; but, in fact, it is the quality of your attention and thought at *every* step that will ultimately determine the success of your planning efforts.

THE GOAL

In a sense, formulating the goal from the challenge is like turning the challenge inside out. If the challenge is the WHY, then the goal becomes the WHAT. It is the pole to jump and clear, the baton to grab, or the tape to break.

Many planners make a distinction between goals and objectives. In this process, goals and objectives are treated as the same thing. One might also call them results. They are what you are shooting for, what you hope will be accomplished when you review your efforts in three months or six months or whatever your implementation window is.

These preliminary goals become guides for your action, sign posts that point you in the right direction to achieve the desired results. If your goals are accurately derived from your challenges, they will assist you and your team in ensuring that any action you take will directly address the top-priority challenges that have been identified.

The most important task to accomplish in setting your goals will be to make sure they are measurable and directly derived from the challenges your team has outlined.

THE STRATEGIC ACTION

Here is where the miracle happens and/or where the fun starts, depending on your point of view!

The strategy that follows the outlined goals is the HOW of the strategic action process. The strategy specifies the action needed to accomplish the goal.

This part of the process is sometimes the most invigorating because, even if there has been agreement on the challenges and goal-setting steps, there is bound to be spirited discussion about how to meet those goals. The truth is, there are many different ways to accomplish the same goal or set of goals.

As an example, say your business is manufacturing and selling toothbrushes. Your challenge is that the business has been losing market share for the past three or four years. Therefore, your team has decided that one of the general goals is to increase sales.

Someone on the team has proposed the idea that you could broaden your sales base by including a new target market—pet owners—by producing a new product—a toothbrush for dogs and cats! He points out that the pet products industry has grown every year for the past decade; that the Baby Boomers are just entering retirement, many will be empty-nesters and their attentions will be even more focused on those four-legged members of the household. And that perhaps with a small change to your current product, your company could be entering a relatively uncompetitive niche.

After the laughter dies down, your team discusses the marketing and manufacturing research that would need to be done to validate the idea. The marketing rep is already thinking about where to get figures for the potential target market. Who is in the market now? Do other comparable products exist? Meanwhile, the sales representatives on your team are brainstorming possible sales strategies. How will you reach this new consumer group—through their regular dentists, veterinarians, or simply by marketing in a centralized way to the national pet store chains? Will they need to hire someone with contacts in the animal world? Or perhaps all of these approaches would need to be combined into a coordinated sales effort, but which one should happen first? Then the PR person jumps in, "But will the image of animal toothbrushes damage our current consumer marketing approach? Maybe we need a new trade name?"

This hypothetical scenario represents the kind of discussion that could happen when your team gets to the strategic action formulation stage. Once a proposed strategic idea is selected, it could lead to the creation of several strategic action steps in your planning document. And, as you can perhaps imagine, especially if you are working as a cross-disciplinary team, there will be many differing opinions about HOW to meet even an agreed-upon goal.

The chapter devoted to strategic action outlines procedures that you and your team can take to begin with an open field of ideas. Often, the challenge in the planning process is not to limit one's options too soon. This might be the point in your process when you get to know most about the individuals on your team. Some team members will want to forge directly to ACTION without spending time doing the exploring and questioning that must be part of any strategic planning process. They will be fed up with the theoretical. Whereas, some of your members will rise to the fore and offer myriad suggestions and ideas for the HOW. It is at this step in the process when all your thinking and understanding about insight comes into play. And *play* is absolutely the right word. Perhaps the pet toothbrush idea will be thrown out as a good joke—but it might also end up being the kind of innovative marketing idea that provides a breakthrough for your company.

At any rate, devising the strategic action is never an endpoint. Even if it outlines a comprehensive idea, it is an idea that must be monitored for efficacy.

MONITORING AND MEASUREMENT

Devising measurements and monitoring activities, also called 'control' in some processes, will be essential to your team's successful completion of your planning process. Measurements outline the terms of your achievement. Without clear measurements, you will have no way of knowing whether you will meet or have met your goal. The monitoring process is about finding out what is really happening in order to know if your plans are on target. Monitoring and measurement devices generally ask questions about WHO, WHAT, and WHEN.

- What will success look like?
- What kind of measurements will be tracked?
- When will the measurements be taken? How often?
- When will the measurements be reported? How often?
- Who will do the tracking computation, analysis, and reporting?
- Who will receive the tracking information? Who needs to know what?
- Who will be responsible for what tasks?
- Who is accountable if tasks are not completed?

The answers to these and other questions will lead you to the appropriate aspects of the measurement and monitoring step in the planning process. It is only by having concrete benchmarks that you will know if strategic actions need to be adjusted mid-course. Maybe the goal itself will prove to have been too easy, too unrealistic, or even irrelevant.

Chapter 9: Monitoring and Measurements discusses the characteristics of control that provide a ruler for your strategic actions. Further, it outlines how to identify key contingencies that will need to be monitored, how to assign accountability, and how to provide for an environment that supports the success of your plan.

This process of setting the measurements for your goals, and, at the same time, monitoring your trajectory for getting to your target, is the final step in this strategic action planning process, although it is not the end of your project.

The post-planning chapters on implementation and evaluation will also provide information vital to the success of your planning project.

SUMMARY: THE PROCESS

One final word about the planning process itself. There is one very important idea to keep in mind throughout the formulation of all your strategic action planning steps: this is a living process. Outlined here is a planning system, a sequence of steps, that delineates the basic tasks for successful completion of a planning project. But keep in mind that you and your team have the latitude to do what is needed to make things work for you. There is a John Lennon song that captures this idea perfectly: "Whatever gets you through the night, it's alright...it's alright."[2]

In other words, make it work for you. Whatever works is right. The planning process that you and your team devise, perhaps based on some of the ideas outlined here, must suit your organization. If your organization values thoughtful research, if everyone likes to be kept informed about project details, if there is a tendency to communicate on paper with multiple cc's, if your decision-making process is by consensus and based on a lot of dialogue, then your planning process might take six months or more and result in a hefty, detailed document.

On the other hand, if your organization utilizes email as its main method of communicating, if your team convenes virtually for reporting-back sessions only and works in small groups off-line, if your boss's attitude is "Give me the last line of your report first," then your planning process may take three weeks and require only a three-page executive summary planning document.

The process must be appropriate for your culture. At each step of the way, you and your team will be creating a framework that is meant to assist you—the human executors—in the completion of a directed task. Do not allow the framework to get in the way of your success.

Another thing to keep in mind is that nothing is forever. Things change. The world is a place that is constantly reinventing itself. And because each of these four steps is intricately linked to every other step, if one changes, all of them will need to be adjusted. As stated before, each step in this process becomes the most critical one while you are working on it.

During the course of the timeframe you have allotted for the completion of your project—a quarter, six months, a year—it will be part of your job as team leader or coach to take the temperature of the project at regular intervals and to make sure that your team efforts are on track.

You may find that some goals were set too low, or that there is no chance to reach others because circumstances have drastically changed. A sudden merger of your company with another might mean that the challenge identified as the entire foundation of your strategic action has disappeared completely and been replaced by something else. Maybe you will find that the strategic action was all wrong.

Remember, too, that your planning process will not progress in neat steps labeled "challenges," "goals," "strategic actions," and "monitoring and measurements." The human mind simply does not work like that. A human is more than a multi-tasking machine—humans are complex organisms with feelings, moods, facts, images, memories, and associative processes constantly at their disposal. What is more, when you put a group of human beings together to communicate with one another, another layer of complex dynamics takes place.

Thus, whether your planning schedule gets blown out of the water or an argument erupts between the marketing rep and the product engineer on your team or whatever happens, I guarantee that you will need to tinker with what you build with your group. The more flexible your process is, the stronger it will be. In the San Francisco Bay Area, we build with wood because a brick wall cracks when the earth shakes.

The bottom line?...be ready for an adventure. Be flexible.

NOTES

1. Fogg, C. Davis, *Team-based Strategic Planning: A Complete Guide to Structuring, Facilitating and Implementing the Process*, American Management Association, New York, 1994, 19.
2. From the album *Imagine*, John Lennon and the Plastic Ono Band, Apple Records, 1971.

II

THE PLANNING PROCESS

5

SEEING THE CHALLENGES

All truths wait in all things.

Walt Whitman[1]

If truth waits in all things, then it is your job as team facilitator to assist in revealing the truth that exists in the structures, ideas, situations, and even personalities around you. Schrage says, "a gifted leader is capable of creating a meeting environment that fosters collaborative conversations and brings the group to greater understanding and focus."[2]

If that sounds like a big job, it is! But not an impossible one.

A critical place to start is in making sure that the parameters of your team process are clear and accepted by all team members. So, just be sure that, before beginning the four-step procedure for planning outlined in these next five chapters, you have already taken your team through Exercise 1: Creating Agreement. This will guide your team in establishing ground rules for your meeting communications, roles and responsibilities, and project commitments.

That done, this planning procedure can now get off the ground.

WHAT IS WRONG WITH THIS PICTURE?

A colleague of mine who manages a welfare office in the Midwest recently made an interesting comment. I had called him because I assumed that all the budget cuts and new policies in social services that Congress had been passing in the last several years meant that the atmosphere in his office must be pretty grim. I assumed that the workload was impossible, that his staff was depressed because of all the families who would now go unserved, and that, certainly, social services was not a place one would want to be working in the late 1990s.

"It's not always easy," he said, "but, in a way, it's just the opposite of what you might think. The spotlight is on us now, it's true. But there's a tremendous amount of energy in my office. Everyone is having to rethink everything."

Suddenly in his office, 'we've always done it that way' was not a good enough answer to the question of why things could not be changed. Everything was being scrutinized; the entire status quo was challenged. Only days or weeks before, everyone had assumed that they must work with the way things had always been done; but now, every aspect of doing business provided a possible opportunity for innovation.

Despite, or perhaps because of, the tremendous challenges laid out before his staff, no spirit of defeat existed. On the contrary, what my friend was experiencing in his office was a surge in energy and inspiration. Change, even enforced change, was providing a focus and renewal of energy, a new beginning, and a chance to redo and rethink systems in a comprehensive way.

Identifying the challenges in U.S. social services agencies took a groundswell of public opinion and Congressional legislation. How can your planning process inspire a similar level of commitment to change and innovation? Can you set the tone and encourage a truly innovative exploration of challenges in your business?

One of the truisms about planning is that it works best when change is necessary. If your business is facing some impending financial crisis, or a competitor has just revolutionized the marketplace by introducing a product using new technology, or you are experiencing either explosive growth or market stagnation, these are clearly situations in which something must be done; the status quo cannot be maintained.

But more than likely, your organization is not facing conditions that force change. A more common situation in planning is that things look OK for the moment. In this situation—where there is no apparent reason to change—you must still look closely, with insight, to see if there are any patterns that exist in your organization that could become problems in the future. In this situation, you and your team should be conducting a preventative planning process.

You may remember those children's puzzles called "What's Wrong with This Picture?". In these drawings of everyday objects and settings, there are usually several things amiss. Someone's sock is mismatched. A picture frame is missing a side. One side of a tree is drawn in front of the fence and the other side behind the fence. The cat has only its left set of whiskers.

This childhood game is actually a pretty good metaphor for "Seeing the Challenges." The game is never as easy as it looks. Your mind's eye is tricked into *not* seeing what is out of place or missing because the mind is looking for coherence and often fills in the gap if there is one.

In the same way, the problems in your business might be masked by things that have existed for so long that they may be impossible to "unsee." The challenges are there, but they are invisible because you are conditioned not to see them; or even if you can see them, the assumption is that they cannot be changed.

Some think that one of the reasons the British Army was defeated during the American Revolution was because British soldiers were accustomed to marching in colors in squadrons; they brought their military traditions with them across the ocean and applied them in a new land where conditions were very different. Our Yankee sharpshooters, battling in the rolling hills and dense forests of Pennsylvania, New Hampshire, Virginia, and Massachusetts, thought it made more sense to shoot from behind the trees; and Redcoats in formation were pretty easy to spot.

Are there traditions in your industry that no longer make sense? Are there processes and procedures that exist just because "that's the way we've always done things?" What are the 'sacred cows' in your corporate culture? Are they still serving the aims of your organization?

IDENTIFYING CHALLENGES: AN OVERVIEW

A planning process properly begins with a situational analysis. There are dozens of strategic planning books that propose dozens of different methodologies for situational analyses. There is a range of procedures from conducting in-depth research and statistical analyses using comparative historical data and projections into the future, to simple survey vehicles. An entire industry of consultants and analyst organizations has grown up around this task.

The first planning step, "Identifying the Challenges," offers a nontechnical, grassroots method for getting your planning process started quickly. One assumes that either you have been given specific strategic direction from your top management or that you have sufficient knowledge of your business and team territory to be able to identify your major challenges without conducting extensive research. Or maybe both.

If you feel that a more in-depth and classic situational analysis is needed, refer to the "Selected Readings," under the section heading "Situational Analysis." There are several excellent books with a special focus on analysis that either give detail about the process, if you want to undertake it yourself, or suggest parameters to consider if you want to hire an analyst group to conduct research for you.

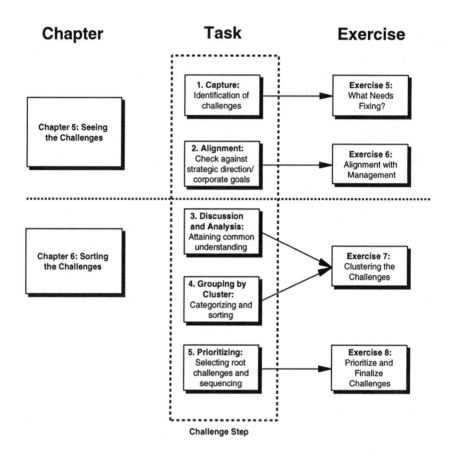

Figure 5.1   Overview of the Challenge Step

As for us, the first step of this strategic planning process is made up of several substeps. It is the most complex of the four major planning steps. The component tasks for this step are outlined in schematic fashion in Figure 5.1.

The process for the entire challenge step contains the following substeps:

1. **Capture:** identifying new challenges, and/or reviewing or updating existing challenges.
2. **Alignment:** checking against strategic direction or corporate goals that are mandated by your executive leadership team.
3. **Discussion and Analysis:** evaluating challenges in order to come to a common understanding of what they are and to reveal their interrelationships.

4. **Grouping by Clusters:** categorizing, sorting, rewording, or refocusing the challenges based on discussion and analysis.
5. **Prioritization:** selecting the key or root challenge in each cluster, arranging them in order of importance or impact, and/or assigning a sequence if necessary.

Just a few words about how this rather more complex series of tasks will be approached. The entire challenge step can be completed in four exercises. There will be one exercise for the initial identification of challenges. A second exercise will assess your team's alignment with top management regarding challenge identification. A third exercise encompasses steps 3 and 4 above: challenge discussion, analysis, and clustering; and a fourth exercise will cover step 5, selecting final challenges.

Steps 1 and 2 will be completed in this chapter. Chapter 6 picks up steps 3 through 5.

TASK 1: CAPTURING CHALLENGES

—identifying new challenges, and/or reviewing or updating existing challenges.

This step begins by you and your team completing an informal exercise called "What Needs Fixing"? In this exercise, your team will generate an initial set of challenges. If you have a strategic mandate that has been given to you by an upper management team, hold that aside for now. Let your team generate its own list of challenges first, and then review the corporate mandates as a check on alignment.

In this exercise, a type of mind mapping technique is used as a starting point. A 'mind map' is simply a free-associative drawing or series of ideas sketched out informally in a combination of words or pictures, sometimes drawn in the style of icons. In this case, that old saying, "a picture is worth a thousand words" is true because a visual image or metaphor can capture succinctly a more complex situation. The exercise uses this technique to lead your team through the process of beginning to 'see' what challenges your company or department is facing.

This mind mapping technique is a way to allow the individuals on your team to be free of the constraints imposed by a more formal discussion. And because it is a mixture of words and images, and something that is privately constructed, sometimes a more creative or intuitive part of the brain can be utilized in the process of capturing information.

Figure 5.2 is a sample of how this technique works. (This is a copy of a mind map that I drew as part of a similar exercise for my own business.)

I started with my business, Axioun, at the center and put a kind of radiant halo around it to indicate the key strengths—some people call these key or core competencies—that the business already has. Then I started indicating the aspects of the business that needed to be strengthened.

First, I drew the file cabinet as a way of symbolizing my desire to create a better system for organizing information. I generate a lot of paper, read tremendous numbers of paper-based materials, and often print information from the Internet as well; I can attest to the myth of the paperless office. I want to create a filing system that is based on my real-world needs.

Second, I drew the BIG PR letters with the muscled arms as a way of inspiring myself to re-evaluate and refocus my own marketing efforts. I drew what I want to exist, not what currently exists.

Then I drew a whole series of ideas in no particular order. I am bringing on a new partner to assist with competitive analysis and research; hence, the stick figure with the helping hands. This is "something that needs fixing" in that I do not want to downplay the effort it will take to incorporate another individual into my organization. That tangle of telecom machinery needs to be put into order and, like the filing problem, tackled with an eye to efficiency both in cost and time. My new venture as an author has a triangle under it, indicating the need to weigh the value of my efforts there against the rest of my business activities. Those hovering, almost invisible, squares of "Poetry" and "Paris" indicate my desire to integrate disparate aspects of my life that are important to me but feel disconnected. I want to find ways of creating synergy between my current business goals and those activities and places that I like but which feel more peripheral than I want them to be.

This is a sample of how mind mapping can be applied for a specific purpose, and it is an example of the kind of mind map you and your teammates might create. Note that I have not shied away from including some personal aspects in my mind mapping, although the focus is on my business; this aspect of the 'whole-self' and the importance of the engagement of the self is discussed further as the book progresses. There is no reason to leave out personal aspirations or desires; these can sometimes provide information that helps teams assign tasks, evaluate relevancy of certain activities, or identify those passionate hot spots for a company or work group. Often, it is these feeling markers that spark planning synchronicity.

Now take a look at the following exercise.

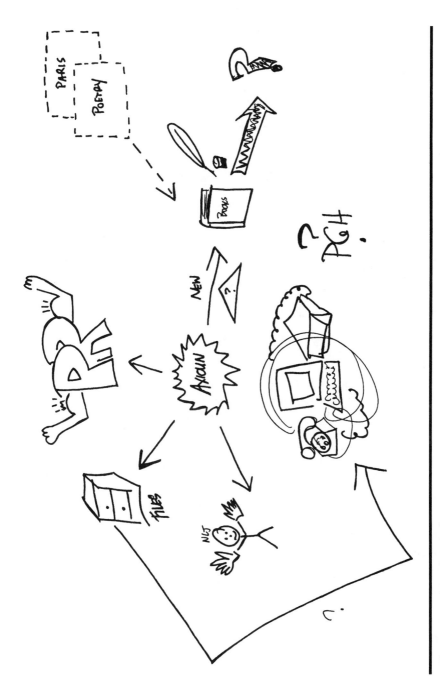

Figure 5.2 Sample Mind Map

EXERCISE 5: WHAT NEEDS FIXING?

Materials: flip charts or large pieces of paper, markers, masking tape.

Duration: 60 minutes.

Objective: to create individual mind maps and discuss them as a group.

Hand out flip chart pages and markers.

Set the exercise up by talking a bit about the child's game, "What's Wrong with This Picture?", and explain what a mind map is. Point out that the pictures should not be fancy. In fact, one can use simple objects to represent complex ideas—like a smiley face or a frowny face to represent your customer base. Emphasize that there is no right answer and that drawing skill is not the point of the exercise!

Ask team members to draw a mind map using words or pictures that illustrate the things that they feel need fixing in your business and/or your department. Point out that their reach does not have to be limited to just the department, nor does it have to be only about your business. They can comment on the business environment or trends they see developing that might endanger the current business strategies. Personal information is also OK.

Ask each team member to work on his/her own map individually. Allow 10 to 15 minutes for this portion of the exercise.

Facilitation tip: During the mind map drawing, keep your eye on what your team is doing. One can generally tell if people are still working or whether everyone is slowing down—when two-thirds of your team looks finished, say something like, "OK, take another minute or two to finish your drawing."

Then bring the team back together and open the discussion to the group. Let those that want to share their drawings tape them on the walls and talk about them.

There will more than likely be some repetition as each team member talks. Do not worry about that—often there are slight variations in how individuals represent idea relationships. You can sort that out later.

When every team member has been given a chance to share his/her ideas, make a quick review of all the mind maps as a group and, on a new flip chart page, record the ideas one by one as accurately as you can using words and phrases.

If something appears more than once, put a check by it for every time it is mentioned.

Facilitation tip: If time does not permit a group discussion and compilation of main points, someone can do this part of the exercise during a break; but if this is necessary, it is important that the list be discussed with the entire group before moving on to the next exercise.

Be sure to get agreement on the phrasing—the translating of challenges from the mind maps into words.[3]

Hang onto this master list; it will be used in a moment.

SAMPLE PLANNING ITEMS: BRAINSTORMING CHALLENGES

For the purposes of this sample procedure, suppose the hypothetical planning team is a marketing department. The team's list of "What Needs Fixing" taken from the mind maps and put into a word list could look something like this:

- ✓✓✓✓ Bad communications with sales department/sales reps: they always get the offer wrong, not prepared for product changes, etc.
- ✓ Lack of follow-through with customer service reps, i.e., "bad press."
- Logo and collateral outdated—need new look.
- ✓✓ Need a 'cooler' Web site.
- Procedures for hiring outside graphics vendors too cumbersome.
- Competitor X is poaching our best customers.
- Department is down two staff members.
- ✓ Support staff not responsive enough to deadlines.
- Why not create a new product for the nostalgic Baby Boomers?
- ✓✓ How can we get more timely competitive info?
- ✓✓✓ Increase sales revenue to meet team goals.
- ✓ Re-evaluate market share numbers.
- Get better coffee and real cream!
- Fax too slow.
- ✓✓ Cut down paperwork!!!!!!
- Benefits for significant others?

✓ checks indicate the number of times a particular challenge was mentioned by a team member.

TASK 2: ALIGNMENT

—checking against strategic direction or corporate goals that are mandated by your executive leadership team.

After the team's initial list of challenges has been generated, review the overall challenge or set of strategic directions that has been delivered to you by upper management—unless your team *is* the upper management; in that case, you can skip this section and go on to the next chapter.

Typical strategic objectives from management might be something like the following:

- Increase revenue by 12% and grow the market share in our three-state region by 8% in the next two years.

- Identify and create awareness for our company in at least two new international markets and four new vertical industries.
- Raise additional capital of $750 million.
- Improve internal quality and reduce our manufacturing defects by putting out 99% defect-free packaging by the end of the year.

Sometimes, top management will also share information about strategic direction as well, with statements such as the following:

- We will continue to compete on price in markets X, Y, and Z.
- We will seek customers in markets A and B that are less price-sensitive and, where we cannot compete on price, offer superior customer service.
- We will undersell Competitor X in overseas markets in order to increase market share.
- We will maintain status quo in sales of product C and explore the feasibility of entering the market with a lower-priced alternative.

Do you see some convergence in the lists of concerns and challenges that have been generated by your team and those that have been given to you by your management team? Is there basic agreement on the major issues? Or is there some obvious discontinuity?

Major disagreements could point to a need for some alignment discussions with key members from your team and your management group. Sometimes, the differences are only a matter of semantics or incomplete understanding of business goals; but it is best to get those cleared up before your planning process gets too far ahead of itself, since one of the jobs of your team is to support corporate objectives.

More ominously, differences between what your team thinks are the major challenges or directions and what upper management thinks could reveal major stumbling blocks to your planning process.

As mentioned previously, if top management is oblivious to changes that must be made in order to maintain the vibrancy of your business or business unit, you probably have a bigger problem than you or your team can tackle alone. On the other hand, if top management is pushing for innovative changes that your team is dragging their feet about, you would be better off suggesting that your people get on the bandwagon.

EXERCISE 6: ALIGNMENT WITH MANAGEMENT

If you sense that there are discontinuities between the challenges your team has identified and those you have been given as strategic direction from your executive team, you might need to conduct the following exercise to determine how to handle the discrepancies.

EXERCISE 6: ALIGNMENT WITH MANAGEMENT

Materials: flip charts and markers, list of challenges from last exercise, statement of management strategic direction or corporate goals.

Duration: 60 minutes.

Objective: to discuss the team's list of challenges and assess alignment with upper management's goals.

The exercise is simply a discussion to review preliminary ideas and direction. Set up the discussion by going over the executive goals. Then go over the list of challenges that has been compiled from your team exercise. Ask team members if they see any discrepancies or if they feel there are any conflicts between the two sets of goals. There might be areas of concurrence and areas of divergence. This can be a politically sensitive discussion—do not be discouraged if it takes some time to prime the pump.

Record potential conflicts or areas of misalignment—perhaps by underlining them in contrasting color. Then explore options for going forward: a discussion with the executive team handled by a smaller group from your planning team; you, as team leader, taking the concerns of the group to one of the executive team members; bringing one of the executives into the next team meeting…your team will have a sense of the best way to proceed.

Even if the objectives received from upper management give you and your team members guidance as to what the business leaders in your organization are prescribing for growth, they do not necessarily provide the tactical or strategic action detail needed to get the results: that detail you and your team will need to provide based on your insider's knowledge of the business, the environment, and your competitors. Translating those broad management objectives into strategic action will come from your team's list of challenges.

Now, you should have completed subtasks 1—capture of challenges— and 2—challenge alignment with upper management. The following chapter picks up where this one leaves off and takes you through the remaining subtasks in this first major planning step.

NOTES

1. Walt Whitman, as quoted in the *Zen Calendar*, Sunday, November 30, 1997.
2. Schrage, Michael, *No More Teams!*, Currency, Doubleday, New York, 1995, 123.
3. *ibid.*, As Schrage says in *No More Teams!*, "Lowell Steele, a former General Electric executive remarks that 'if you want to control a meeting, get to the

blackboard first...'" (p. 133). Whoever gets to the whiteboard first can control the meeting because it is language that gets written down and the person who controls that holds the power. So be careful about your use of language when you are attempting to capture ideas. Be clear and complete. Make sure you have agreement on the wording and meaning as you go along.

6

SORTING THE CHALLENGES

...the part of this world that we can inspect and analyze is always finite. We always have to say the rest of the world does not influence this part, and it is never true. The world is totally connected.

Jacob Bronowski[1]

CONNECTION AND DISSECTION

Would you guess that chloroform could have an effect on the high-tech industry? Guess again. Scientists are working on a new computer that has a processor made of hydrogen and chlorine atoms.[2] Are the Ecuadorian jungles connected to the pharmaceutical industry? Of course. Many patents have emerged from Amazon forest substances that native peoples have known about for generations.[3] Could a corporate strategic planner learn anything from the story of the Trojan War? I think so!

If the world is totally connected, then information from any part of the world is relevant for any other part; it is simply a matter of how skillfully or creatively one puts the information together; or how clearly one is able to see the connections. Seeing connections is a talent that artists, poets, musicians, inventors, architects, children, and strategists share. It is an important aspect of strategic planning.

But at the same time that one acknowledges the importance of being able to connect,[4] one must also discuss the value of being able to dissect. Analysis is a process of separating "an intellectual or substantial whole into its constituent parts for individual study."[5] The word comes from the Greek *analusis*, a dissolving, or from *analuein*, to undo.

Your team has identified challenges, and the next task will be to understand the nature of those challenges. What elements are they made up of? What are they caused by? How are they related to one another? This chapter is about how to take information apart.

Separating and sorting the information of the world—or sets of worlds—that we live in are critical steps in understanding. This chapter examines how to analyze and make sense of the challenges that have been identified by your team. The subtasks in this phase of the challenge step have to do with sorting the challenges into clusters—separating them out from the whole and at the same time seeing the connections between them—and selecting the main, or what can be called the "root," challenge in each cluster.

As a reminder, we have completed substeps 1 and 2 in the whole challenge identification process. (These are outlined in Chapter 5.) Substeps 3, 4, and 5 remain:

3. **Discussion and Analysis**: evaluating challenges in order to come to a common understanding of what they are and to reveal their interrelationships.
4. **Grouping by Clusters**: categorizing, sorting, rewording, or refocusing the challenges based on discussion and analysis.
5. **Prioritization**: selecting the key or root challenge in each cluster; arranging them in order of importance or impact; and/or assigning a sequence if necessary.

These tasks will be discussed first, in order to provide a methodology for leading your team meetings. Then the task itself will be outlined in two exercises. Exercise 7: Clustering the Challenges involves discussing the challenges identified by your team and grouping them into idea families or clusters. Exercise 8: Prioritize and Finalizing Challenges proposes a final check on the relevancy and format of your challenge list.

TASK 3: DISCUSSION AND ANALYSIS

—evaluating challenges in order to come to a common understanding of what they are and to reveal their interrelationships.

Generating challenges individually has allowed each person on your team to begin the process by delivering personal input. This is important, both for buy-in to the process and because you want to throw as wide a net as you can in these initial stages of capturing your challenges. Now you need to begin the process of directing and facilitating the dialogue with your team to reach a common understanding about exactly what those challenges are.

Remember, however, that at every step of the way, it is the process that is your most important product. True, you need to generate information. Ultimately, you need words on a page, a plan, some measurements, some deadlines—but, in a very real sense, these are byproducts of the process itself. If you establish early on a structure that allows the process center stage, those words on the page will simply be a road map for destinations and locations that everyone already knows. Yes, you need the road map in order to keep reminding yourself what you said you wanted to accomplish, but building trust as a team and creating a common understanding of where you are going and why is the real work.

So, although some suggestions about the way to manipulate the challenges as content are offered here, remember that the process and the quality of your discussion are what count.

OK, how are you going to make sense of your preliminary challenge list? Here are some questions to ask yourselves as you think more deeply about the list of items generated.

- Do any of these challenges seem related to one another? Which ones? How are they related?
- Are any of these challenges causal to any of the others?
- Are any of these challenges the result of another problem? If so, what is the source problem?
- Do any of these challenges have a common problem source or *root challenge*?
- Are we really describing a problem/challenge, or have we identified the symptoms for something else?

You might want to begin this portion of the discussion by posing these questions to your team. Then take a look at each of the preliminary challenges on your master list. Talk about each one separately. Do not proceed too quickly; try to solicit opinions from the team members who seem quiet—often they have a different opinion or another idea about something and just need a little encouragement to speak out. Use your intuition as you lead the discussion. Sometimes you may feel discord just under the surface that people seem unwilling to talk about. Try to solicit it. Or, if you sense general agreement, push the group forward by restating the challenge in a way that captures the consensus. Then move on to the next item.

In the course of this discussion, there may be other challenges that surface. If so, keep adding these to your list. Do not throw anything out yet. If one challenge seems related to another, draw an arrow.

Look at the sample challenges again to get an idea of how the discussion might go:

- √√√√ Bad communications with sales department/sales reps: they always get the offer wrong, not prepared for product changes, etc.
- √ Lack of follow-through with customer service reps, i.e., "bad press."
- Logo and collateral outdated—need new look.
- √√ Need a "cooler" Web site.
- Procedures for hiring outside graphics vendors too cumbersome.
- Competitor X is poaching our best customers.
- Department is down two staff members.
- √ Support staff not responsive enough to deadlines.
- Why not create a new product for the nostalgic Baby Boomers?
- √√ How can we get more timely competitive info?
- √√√ Increase sales revenue to meet team goals.
- √ Re-evaluate market share numbers.
- Get better coffee and real cream!
- Fax too slow.
- √√ Cut down paperwork!!!!!!
- Benefits for significant others?

√ checks indicate the number of times a particular challenge was mentioned by a team member.

A discussion of these challenges might reveal, for instance, that poor communications with the sales reps is due to the headquarters vs. branch separation; and the fact that there are multiple hand-offs of information—first to sales managers and then to sales reps. Maybe the reps are too far removed from the source of information. These challenges sound as if they are related by the root challenge or problem of "poor communications between headquarters and branches."

Someone might offer that there is no good email in the branches because the IT people are focusing on the mainframe changes needed for the year 2000 turn-over and cannot focus on intranet issues. This might be related to why there is not sufficient manpower to adequately research and update the Web site. Another team member might bring up that more and more customers are requesting email addresses and keep asking, "When is there going to be a Web site for customer FAQs or even a live chatroom?"

During this kind of discussion, your team might uncover other relevant challenges; you may want to restate some that appeared on your original list; or you may end up with questions your department cannot answer.

Probably the most difficult aspect of this discussion will be keeping your team focused on exploring your challenges and not lapsing immediately into problem-solving, especially for the more results-oriented members of your team who will want to jump ahead immediately into solutions.

Keep reinforcing the idea that you are exploring together for the purposes of understanding and particularly for an understanding of issues that may at first be hidden from view. Take your time. Remember, you can recognize strategies because they deal with HOW—you are still dealing with the WHYs and the WHATs.

If you have discussed each challenge thoroughly and your team has a common understanding of the challenges, proceed to the next step. These two steps are obviously related and often naturally happen at the same time or flow one into the other. In fact, Exercise 7: Clustering the Challenges assumes that you will lead your team in both discussing and clustering tasks at one time.

TASK 4: GROUPING BY CLUSTERS

—categorizing, sorting, rewording or refocusing the challenges based on discussion and analysis.

This technique is called *clustering*. It is a method of grouping items derived from like ideas into systems or families. The next step in the process of analyzing your challenges is to sort them by type as a way of:

- Making sure you have not missed anything
- Identifying the sphere of influence that they operate within or are derived from, or identifying the 'system' they are part of
- Grouping like challenges together in clusters to better understand how to handle them and who might handle them

Often, aspects of a situation that are troublesome are part of a larger, more complex system—it is generally not just one thing that is amiss—and by looking at the whole picture, your team will be able to get a better view of what can be accomplished and how to make needed changes. It is a tactic that will help during the strategic action step in your process.

Your team's brainstorming on challenges is really a grassroots situational analysis. There are many ways to organize information in order to make sense of it. One that might be helpful is to take a traditional approach

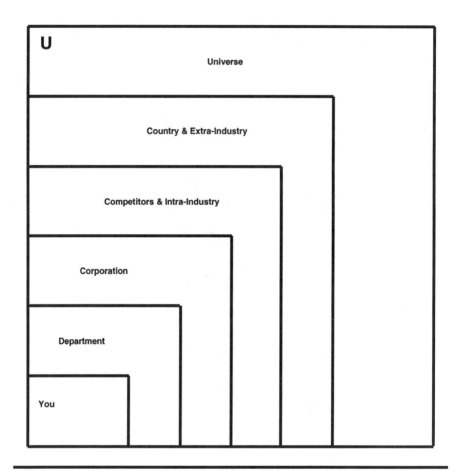

Figure 6.1 Organizing the Universe into Internal and External Influences

that divides the world into internal and external to the company. Figure 6.1 illustrates this system.

This chapter opened with the idea that everything is connected, so there is one important thing to remember about diagrams like Figure 6.1: this, in effect, is a systematized description of the world; and since the world is only and always itself and our organizational system artificially separates all of its natural and complex connections, this representation is, perforce, faulty. It requires a simplification that is contrived, and the more complexity that is taken out of a system in the process of codifying it, the less the resulting code is able to represent reality accurately. The short version is that not everything will fit neatly into one category.

An elaboration on the earlier quote by Jacob Bronowski phrases this dilemma superbly:[6]

I believe that there are no events anywhere in the universe which are not tied to every other event in the universe.

—the part of this world that we can inspect and analyze is always finite. We always have to say the rest of the world does not influence this part, and it is never true. We merely make a temporary invention which covers that part of the world accessible to us at the moment.

Nonetheless, for purposes of your analysis—any analysis—you need some kind of organizational system in order to deconstruct and therefore understand the focus of your inquiry; so make do with one.

One can define the layers of this inside-vs.-outside view of the world in a top-down order as follows:

- Universe: economic, environmental, political, social, cultural, and technological trends worldwide.
- Country and extra-industry: pressures, changes in the environment, general lifestyle changes, fads, technology, and any industry trends that are affecting your country or continent.
- Competitors and inter-industry: competition, new market entrants, competitors' strengths and weaknesses, niche technology, and industry-specific trends and influences.
- Corporate/internal: business-wide, corporate culture and values, any policy applicable across all departments (hiring policy, benefits, etc.), management style, management hierarchy, and financial health.
- Division or department specific: department personnel, department functionalities, department policies, structure, and culture.

Assumably, your strategic direction has been established by your upper management team; probably your strategic challenges will fall into the last three categories—competitors and inter-industry, corporate, and departmental—which comprise the lower half of the world view. Of course, whether or not this is true will depend on the planning process in your organization, what level of decision-makers are involved, and what the boundaries of your team's decision-making are.

Again, the real world is infinitely richer than these categorical boxes. And because of that, one needs to remember that these categories overlap and influence one another: there are many challenges that link to others or are supporting challenges for other issues in other categories.

This inside-vs.-outside view of the world is simply one way of organizing things; you and your team might devise another world view that makes more sense than this one. If your business functions primarily in project team groups, or cross-functional teams, these traditional ways of looking at a business might not be helpful. You may need to create new buckets for sorting your challenges into client groupings, or products and services, or geographical regions. Whatever helps you and your team to analyze and understand your challenges is the "right" way to do it.

The following exercise provides direction for leading your team through the two steps outlined and discussed above.

EXERCISE 7: CLUSTERING THE CHALLENGES

Materials: flip charts and markers, all challenges written on visible flip charts and posted on meeting room walls or handed out as a list.

Duration: 60–75 minutes (if the meeting runs longer, schedule a second session).

Objective: to discuss and evaluate your preliminary challenges in order to come to a common understanding of what they are; to categorize or cluster the challenges into family groupings; and to choose the "root" or main challenge in each group.

Begin the meeting by outlining the meeting objectives and the specific tasks above. You may want to pose to the team some of the discussion questions that were suggested on page 77.

The best way to proceed is to go through all the challenges one by one and simply restate what the challenge is. If other items arise in the course of this discussion, write them down on additional flip charts. Or if some items now seem repetitive, cross one out.

Facilitation tip: Make sure that all of your challenges are written on flip charts and posted so that the entire team can see them and everything is visible at once. This helps people see the connections between items and can often provide visual cues about how ideas can be combined or might be related.

Facilitation tip: Do not let the discussion get bogged down on issues that should be discussed off-line, either with a smaller group of people or with someone who may not be present at the moment. You might want to have another flip chart sheet devoted to "Questions" or "Things to Check On Later"—something that allows you to take note of a good idea or something you do not want to forget about without letting it sidetrack the discussion.

If it is feasible—especially if the group needs an answer to a question in order to continue the discussion—have someone make some calls during the next break and see if any information can be gathered that way.

Once you feel that your team has a common understanding of the challenges, proceed to the clustering task.

You will need to decide how entensively to explain the task of 'clustering' and/or whether a short discussion of an external vs. internal worldview is needed to assist team members in this task.

Facilitation tip: one easy way to cluster challenges is to designate a number or letter sequence and simply assign one number/letter to each individual challenge by writing it on the posted flip charts beside each item. The challenges that share the same number/letter can then be grouped after the meeting and word-processed in a more readable format.

Sometimes, hybrid categories emerge or an item might appear to belong to two categories. Do not worry about fitting everything into a glove-tight system for now. The more informational detail you capture on your flip charts, the easier the job of word processing will be.

As always, assign a team member to draft a document containing the information that results from this meeting. This document will provide the raw material for the next task.

SAMPLE PLANNING ITEMS: CHALLENGE CLUSTERS

After a team discussion and clustering session like the one suggested in Exercise 7, the preliminary list of sample challenges might begin to look something like this:

External Business Challenge/Marketing

- Need new direct marketing campaign to increase visibility in target region.
- Do not know potential new target audiences within our region.
- What do our current customers like or dislike about our product or service? What about new product offerings?
- New collateral to support direct marketing campaign—budget?
- Why are customers leaving for Competitor X?
- Increase our market share in our three-state region by 8% in 2 years.
- ⇓ Do we need a better Web site? including customer FAQs or chatroom?
- ⇓ We need better competitive information at more timely intervals.

Internal Business-Wide Challenge/Communications Issues

- Sales department/sales reps too removed from marketing information, need to hear it from "the horse's mouth?"
- Customer service reps overwhelmed by a rash of 'new customer' problems during our campaigns.

- ⇑ Do we need a better intranet Web site? including customer database access, current sales info, etc.?
- Inadequate email system with branches? realtime, not batch, processing.
- ⇓ Do we need so much paper communication internally?
- Benefits for live-in partners possible?

Internal-Department Challenge/Administration

- ⇑ Contracting procedure too cumbersome? Check with legal.
- ⇑ Do we need so much paper communication internally?
- Need faster fax.
- ~~Better coffee~~ solved!

Note that the arrows indicate that a challenge is repeated in the category either just above or just below it in this hierarchy, meaning that it will probably need to be 'solved' or addressed on multiple levels.

Or if your world makes better sense divided into team groupings or functionalities, your sorted challenge list might look something more like this:

IT/Techie Team

- ⇓ Do we need a better Web site? internal: including customer database access, current sales info (see administration); and external: customer FAQs, chatroom, (see Product X team).
- ⇓ Inadequate email system with branches: need realtime, not batch, processing, (see administration).

Administration/Support Staff

- Contracting procedure too cumbersome.
- ⇑ Do we need so much paper communication internally? (see IT.)

Product X Team

- Do not know potential new target audiences within our region.
- ⇑ What do our current customers like or dislike about our product or service? What about new product offerings? (see IT.)
- Why are customers leaving for Competitor X?
- Increase our market share in our three-state region by 8% in 2 years for sales of X.

Creative

- New collateral to support direct marketing campaign—budget?
- Need new direct marketing campaign to increase visibility in target region.

All/Special Projects

- Benefits for live-in partners possible?
- ~~Better coffee~~ solved!

Note ⇑ or ⇓ arrows indicate a challenge either supports or depends on another team's challenge.

TASK 5: PRIORITIZATION AND SELECTION OF KEY CHALLENGES

—selecting the key or root challenge in each cluster, arranging these in order of importance or impact; and/or assigning a sequence if necessary.

Prioritizing has to do with balancing, weighing, and comparing.

Now that your team has a common understanding about what is meant by the challenges and they are roughly sorted into groupings, the next step involves prioritizing or selecting the key challenge or "root challenge" in each challenge cluster. Keep track of all the challenges within each cluster because, as you will see in the strategic action step, these challenges will provide the basis for the action ideas. And, as mentioned before, the more informational detail you keep as you progress, the better your ultimate planning product should be.

Because there are probably too many challenges or challenge clusters to tackle in one planning cycle, some decisions need to be made at this step about what will be the best use of the team's resources. How many challenges your organization can handle depends on what your timeframe for action is, how many people will be on the implementation teams, and how much of their implementors' time will be spent on these tasks. These are questions only you and your team can answer.

In my experience with planning projects, I offer as a general rule that from two to five goals are about the maximum that can be accomplished well by one team or department in one planning cycle. Three is probably optimum. Attempting to handle more than that often means that team focus and determination becomes too diffuse. The momentum that will be built up in the course of your planning together will be lost.

As discussed before, challenges usually exist in systems or clusters; so you could consider taking on an entire cluster and working on the grouping of challenges over a specific, perhaps longer, timeframe. In the next step—Goal Setting—there is another opportunity to review the challenge groupings and determine if there are too many or too few goals.

The criteria I suggest for your prioritization step are the following:

- Impact vs. Resource Analysis: Would the solution to this problem further our corporate challenge? If so, how? Will the solution for this challenge require a high level of resources? Will solving this challenge be an efficient use of resources per risk/reward?
- Probability for Success: Are these challenges within our power to affect? (If yes, proceed. If no, talk to boss and or take other appropriate action.) Are these challenges appropriate for our team? Do we have the skills needed to meet them? Which challenges have the greatest probability for success?
- Sequencing: Which challenges need to be solved first? What are the dependencies between challenges? Does one challenge lead naturally into another?

These criteria and some ways to apply them are discussed below.

Impact vs. Resource Analysis

To illustrate this criterion, a visual technique adaptable to many different types of business analyses can be used (Figure 6.2). This model employs x-y axes forming a box with four quadrants. One can then compare the most propitious conjunction of variables in order to identify a "sweet spot" or target region within the defined quadrants.

In our example the y-axis is "impact;" that is, impact on the bottom line. Sometimes, depending on the strategic direction of the company, impact can actually be measured in hard numbers; sometimes it becomes simply a judgment call. In the sample planning scenario being used, if the corporate objective is, "We will increase our revenue by 12% and increase our market share in our three-state region by 8% in the next two years," your team might ask, "Does solving this challenge have a high or low impact on revenue or market share?"

The x-axis becomes resources needed to solve the challenge. At this point, since the measures the analysis has not been done, this is a ballpark guess. But, again, given what you all know about your business and

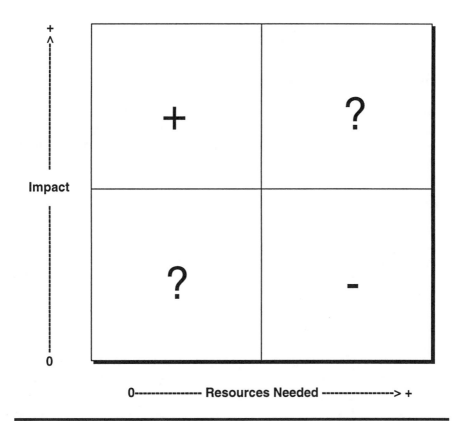

Figure 6.2 Impact vs. Resources Cube

business practices in general, you will probably have a pretty good guess about whether a challenge will use a lot of resources.

You may want to spend a minute or two talking with your group about what you mean by resources. Resources could be time spent in actual labor, time needed to acquire necessary new skills to accomplish a task, or time for training staff to maintain a new system. Resources could also mean money for material expenses, product development, or the hiring of specialized staff to do a part of the task that cannot be done by a team member. When everyone is clear about the definition for resources, you can then ask, "Does the solution for this challenge require a high or low level of resources?"

When you map your challenges against these two qualities, you will be able to see immediately which challenges give you the highest level of return for the energy expenditure. As illustrated, the upper left quadrant is the region of high impact/low resources—the best of both worlds. The lower right quadrant contains the low impact/high resources challenges—stay away from those.

The question marks in the low impact/low resources (LI/LR) and the high impact/high resources (HI/HR) quadrants are there because these are harder calls and probably can only be made based on the long-term direction and goals of your company. The LI/LR may be good risks to take because the investment is low and perhaps the impact will change. On the other hand, the HI/HR may make sense if they open new territory or are worth the expenditures in the long run.

Probability for Success

This selection criterion is definitely a judgment call, although there are a few things to think about in trying to assign it a value. First of all, is the challenge in your power to affect? If yes, then it makes sense to proceed. But if the answer is no, unless there are other very strong reasons for picking this battle to fight, I would suggest dropping it.

In our sample scenario, for instance, "Benefits for live-in partners" is probably not a challenge that a marketing department could successfully undertake. It is very difficult to succeed in changing something one has no real authority over. This challenge or idea for change could be handed off to the Human Resources department as a suggestion from your team.

Some other questions to consider here are, "Is this challenge appropriate for our team?" or "Do we have the skills needed to meet this challenge?" In the first case, if there are a number of team members who feel strongly about, even personally committed to, solving a particular problem, they will be more likely to be effective. In the second case—something that will be reflected also in the resources question—if your team has the skills needed to solve a challenge that appears on your list, then perhaps you are the best people in your organization to do it.

There is another way to look at this success coefficient as well. Does the spirit of this challenge match or fit with the spirit or dynamic of your team? If, for example, your team is noted for its innovative and fleet-footed solutions to problems or solutions that involve sometimes radical change, and you have a problem that looks as if it will take that kind of approach, then that is a challenge suited to your team; whereas, if a problem seems like it will need a slow, steady effort and require lots of detailed follow-up, that is probably a challenge your team will be less effective in tackling.

At any rate, try to give each challenge a Probability for Success quotient from 1 to 10—10 being something with the highest chance for success.

In my experience, despite this seemingly unscientific approach, most teams will be able to come to a consensus about this measurement.

Sequencing

You may have already discussed sequencing when you took up your initial analysis of the challenge list, but it might be worth revisiting. Which challenges need to be solved first? Do you see any other dependencies between challenges? Or does one challenge lead naturally into another?

There is a clear example of this in the sample challenges. Obviously, the question of updated logo and collateral material needs to be decided before another direct mail campaign can be rolled out. These challenges could coexist in parallel for awhile as other problems are being solved; but at some point, the direct mail campaign is dependent on the completion of the logo and artwork decision. This is a sequencing relationship that needs to be identified; this sequencing becomes even more critical if dependent challenges end up being implemented by different groups of people.

The sequencing task is a simple matter of evaluating the chronology of and interrelationships between your challenges.

The following exercise sets the tasks outlined above into a meeting format for your team.

EXERCISE 8: PRIORITIZE AND FINALIZE CHALLENGES

Materials: flip charts and markers, updated and reformatted list of clustered challenges from the last exercise.

Duration: 60–75 minutes (if your session runs longer, schedule a second part at a different time).

Objective: to select the "root" or key challenge in each cluster; to arrange key challenges in order of importance or impact; to discuss probability for success; and to assign a sequence if needed.

Outline the general tasks for the meeting session. For the impact exercise, you will need to explain the Impact vs. Resources analysis tool. Then break the team into smaller groups of two to four people and assign each group the task of applying the impact cube to the list of challenges. Ask each team to actually draw an impact cube and place their challenges in appropriate places on the cube. Allow each team to report back to the large group session, explaining the results of their efforts.

Discuss the overall results. Is there coherence in the results? In what areas do the results diverge? See if the group can approach a consensus on the results.

Perhaps one impact cube can be drawn with the consensus placement of challenges.

Next, discuss each of the challenges in terms of probability for success:

■ Are these challenges within our power to affect? (If yes, proceed. If no, talk to boss and/or take other appropriate action.)
■ Are these challenges appropriate for our team?
■ Do we have the skills needed to meet them?
■ Which ones have the greatest probability of success for us?

This discussion might eliminate some of the challenges. Or, as is often the case, new challenges could be identified.

Last, consider sequencing:

■ Which challenges need to be solved first?
■ What are the dependencies between challenges?
■ Does one challenge lead naturally into another?

Take notes on the results of your discussion and, again, designate someone to do the reformatting and word processing in preparation for the next team session.

SAMPLE PLANNING ITEMS: FINAL CHALLENGES

After the prioritization discussions sparked by the above exercise, the sample challenges might look something like the following. What might be considered the main or root challenge in each cluster has been elevated to the top and put into a box.

Keep the challenge clusters—or you might say overall challenge and subchallenges—together at this point. It is not necessary to delete any of these yet; they will help in formulating both the actual goals and the strategic actions in later steps.

A. Overall external business challenge: increase our market share in our three-state region by 8% in 2 years.

Supporting Marketing Challenges

1. Need competitive/customer info: why are customers leaving for Competitor X? Do not know what our current customers like or dislike about our products and services.
2. Do we need new product offerings (i.e., to take advantage of the aging Baby Boomer market)?
3. Not taking advantage of potential new target audience within region.

4. Current marketing collateral needs new look—budget?
5. Design new direct mail campaign? Other distribution vehicles?

B. Internal business challenge: improve our communications with other, linked departments, i.e., sales, customer service, fulfillment.

Supporting Communications Challenges

1. Sales department/sales reps too removed from marketing information.
2. Customer service reps overwhelmed by a rash of 'new customer' problems during our campaigns.
3. Do we need an intranet Web site for internal customer database and sales info? Or extranet Web site as new distribution channel? (Marketing tie-in above?)
4. Slow or no email possible between headquarters and branches.

C. Department challenge: department communications inefficient (too paper-based) and contracting not expedient (takes too much time).

Supporting Administration Challenges

1. Contracting procedure too cumbersome—check with legal?
2. Too much paper in our internal communications process? Or how do we want to communicate with one another: paper, electronic/digital media, face-to-face?

SUMMARY

Do not be discouraged if your process is not as neat and tidy as these sample challenges appear to be. Any process with living, breathing humans is going to be more unpredictable—and probably more creative and insightful—than this set of linear steps can possibly indicate.

The most important aspects of this first step—Identifying Challenges—is that you work as a team to generate a list of challenges, discuss them carefully and thoroughly, and select and put them together in some kind of groupings. You should end up with between two and five challenge clusters.

Chapter 7 refines the list even further by transforming these key challenges into goals. And, even in this next step of setting goals, the list of challenges will continue to undergo changes.

NOTES

1. Bronowski, Jacob, *The Origins of Knowledge and Imagination*, Yale University, Mrs. Hepsa Ely Silliman Memorial Lectures, delivered 1908; published by Yale University Press, New Haven and London, 1978, 96.
2. *New York Times*, article by John Markoff, "Quantum Leap From Theory to a Powerful Potential," April 28, 1998.
3. See information about Nicole Maxwell, or check out her book, *Witch Doctor's Apprentice*, Citadel Press, New York, 1990.
4. It is really impossible to talk about connections in the world without citing E. M. Forester's brilliant novel *Howard's End*, Signet, New York, 1992, whose moral imperative is "only connect." This is a story set in turn-of-the-century London that chronicles the victories and missteps of a series of lively characters attempting to make connections: in their feelings; in their views on race and class differences; and in their personal relationships.
5. Information taken from the *American Heritage Dictionary*, CD version, produced by WordStar International, licensed from Houghton Mifflin Company, copyright 1993.
6. Bronowski, Jacob, *ibid.*, p. 58 and 96.

7

SETTING THE GOALS

Everything you can do it don't mean it's a good idea, said Boyd.

Cormac McCarthy[1]

WHAT MAKES A GOOD GOAL?

Before the controversy that swirled around United Way in the early 1990s, a pre-holiday fixture in many corporate lobbies was a huge thermometer with an amount in dollars written in bold letters at the top. The bulb of the thermometer was colored in red and the level of red crept up toward the top of the thermometer in increments posted every day or every week during these ubiquitous campaigns.

Sometimes, every department had its own bulletin board thermometer registering its level of giving with a dollar goal posted at the top of the bulb. These bulb totals all fed into the big thermometer in the lobby.

As the red line crept up the corporate thermometer, memos of encouragement and reminders about the approaching deadline were posted. When, finally but inevitably, that red line hit the top, decorations appeared in the lobby, announcements were made, and the "Big Check" made a ceremonial appearance, handed from one person to another, providing the standard photo-op.

Whoever had been tapped and given the task of running the United Way Campaign for the company had been given a packet of information complete with forms, formulae, and directions about how to run the campaign. One thing that was always clear to me in reading over these materials—I was a United Way Chairman in two different business settings—was how systematized and effective they were.

Every kit had lots of tips about the "Giving Goal for Your Organization" and every kit suggested a big visual aid that captured the desired goal and allowed progress toward it to be seen easily and immediately by everyone in the organization.

The United Way's codification of the idea of a goal—how one tracks it and communicates that tracking information to a large and, one hopes, concerned audience—captures brilliantly many of salient features of successful goal-setting.

So, before proceeding to transform your team's challenges into goals, take a look at what qualities make for an effective goal.

Perhaps I should start by saying that, in some planning circles, there is still controversy about the difference between the goals, objectives, tactics, purposes, and other designations for what it is one is aiming for. Although I am a firm believer in the accurate use of language, for purposes here, all of these words will be used interchangeably. They have a variety of closely related meanings, all of which are relevant to our task:

- The purpose toward which an endeavor is directed
- What one intends to do or achieve
- An aim that guides action
- An end result

The meaning of "goal" has its historical roots in sports as "gol," which was a word for boundary or barrier and has come to mean something that a ball gets kicked over. "Objective" is derived from a word that originally meant a 'thing put before the mind.' And the root of "intention" is old French or Latin meaning "to direct attention."[2]

All of these ideas come together to form the modern concept of a goal or objective: something we strive for, something we put our mental attention on in order to achieve. Of course there are different shades of meaning in all the various words one can use for goal—*intent, purpose, end, aim, objective.*

Intent implies deliberateness, while purpose is strengthened by the idea of resolution or determination. In some contexts, goal can suggest an idealistic or even a remote purpose: something one strives for but does not really expect to reach. Aim, because it also has the meaning of aiming toward something, like an arrow shot from a bow, describes the direction of one's efforts but indicates some possibility that the result will not be met—"I'm aiming to leave by mid-May; but if the house does not sell, we'll have to be flexible."

But the point is to choose the word that your team feels comfortable with, or use any or all of them, and do not spend a lot of time arguing about the differences. Goal is used here to encompass all of the possible attributable meanings.

Figure 7.1 Aspects of an Effective Goal

The qualities of an effective goal (Figure 7.1) are probably less controversial than what we call it. Well-stated and effective goals share the following characteristics:

■ Relevant
■ Agreed to
■ Understandable
■ Reachable
■ Quantifiable
■ Assignable
■ Visible
■ Flexible
■ Inspirational

Relevance is another way of stating that a goal is important, even critical, for your organization or department. Chapter 6 offered a few analysis techniques to help you and your team determine the relevancy of your challenge—the Impact vs. Resource Model, for example—so, that aspect is covered. Still, it is worth mentioning again here in this final review. Whatever "makes the cut" and becomes a goal will get passed on to step three for strategic action. And because the spotlight of effort and resources will be strongly applied to all the goals, it might be best to get rid of or reconsider any that you may now feel are not what you want to be spending time and energy on. Or if your team feels that there has been some important directive that was omitted from the earlier discussion, now is the time to include it.

The aspect of **Agreed to** is a comment not only on your goal-setting process, but on your entire planning project. There is no stronger goal than one arrived at by consensus among the people who will actually be working to accomplish it.

If, in the course of your discussion about goals, you sense there is disagreement or misunderstanding about a particular goal, the way it is stated, or the way it will be quantified, stop immediately and get to the bottom of the problem. This might be difficult to do because there will be many team members who are impatient to get on with the action steps—but those seeds of dissent will only grow if they are not addressed. Often, routing out the dissenting view can lead to revelations and information that has not surfaced before, and sometimes this new information can make a real difference in your planning outcomes.

"Agreed to" also means that implementation team members agree to the meaning of the goal and its importance. The common understanding that you unearthed in the process of discussing your challenges one by one should set the stage for the transformation of that challenge or challenge cluster into goals that your team agrees are important and doable. If this is not the case, it is best to start again and discover what the reservations are all about. Better now than four months later, when the goal is still unrealized because team members did not agree with it in the first place.

Understandable in this context means that the goal is clear. Perhaps, in your planning project, the team that has been participating in the identification of your challenges will continue on to devise and carry out the strategic action for each goal; or maybe these goals will now be handed off to a different group of people. But in either case, it is critical that your goal be clearly written and understandable now that it will exist independent of much of the discussion that has preceded it. Here is a place where accurate phraseology can make a tremendous difference.

A goal needs to embody as much of its heritage as it can and still be portable. It becomes a kind of haiku in which every word is dense with meaning, associations, and connotations, so that the complexity of the discussion and understanding that has gone into your challenge identification process can be carried forward into the strategic action phase via this goal.

Reachable is a term used here to describe the right amount of stretch. A goal that is not realistic will only demoralize those who will be mandated to accomplish it. They will give up before they begin. On the other hand, a goal that is too easy to reach, or a goal that is a sham, can have the same negative effect.

Once in a planning session I was part of, we were discussing sales goals for the coming year and mapping them out by quarter. The meeting was taking place just before the holidays, and, as you can imagine, there were many people in our meeting who wanted to be somewhere else.

When the discussion about the first quarter sales goals began, the director of the sales unit was queried about what he felt their booked accounts figure should be. He proposed a number that looked about right and it was written on our whiteboard. Then a few people in the back of the room started snickering, and the facilitator had to stop and find out what was going on. As it turned out, the sales director was more than willing to admit that he had sandbagged his number. He knew he had four or five deals about to close that he could easily book when everyone returned from the holidays. In effect, it meant that, pending some bad surprises, the first quarter goal for his unit was already completed—not a goal, therefore, that would be inspiring much additional hustle.

After some good-natured ribbing, he increased both the figure for the first quarter and the annual number of booked accounts, and, I do not know whether it was just that everyone wanted to get out of that meeting or not, but the sales unit OK'd the new higher goals. Later, they exceeded them as well.

So the question is: what is the right amount of stretch? Past performance can give an indication of standard and achievable goal increases. But the need for growth and productivity can put pressure to increase what has been accomplished in the past. Generally, a balance is reached between what top management would like to see and what the managers and unit members feel is a fair and doable request.

Quantifiable—can the goal be measured? If a goal cannot be measured in some way, you will have a difficult time determining if it has been met. This is a partner idea to the reachable characteristic above. If a goal is reachable, that means you will know when it has been reached; therefore, it must have a means of measurement attached to it.

If goals are quantifiable in nature, it is easier to decide what will be measured: number of accounts booked, percentage of sales growth, return on investment, market share increase—although even these kinds of numbers can be tricky. You will still need to come to an agreement about what exactly is meant by an 8% growth in sales. What is your baseline measurement? In what increments will you be making measurements? At what time intervals? Are you counting booked accounts only or revenue from accounts?

All one needs to do is look at the myriad ways credit card companies figure an interest calculation, for example, to understand that numbers can be manipulated to paint many pictures. Be specific about the one you expect to see.

If goals are less quantifiable—for example, "We want to raise customer satisfaction in our retail outlets" or "We will strengthen our marketing and PR image in the northwestern states"—you still need to establish some terms for monitoring accomplishment. Will you conduct a baseline survey now and follow up six months later with another? Establishing a means of quantifying these kinds of goals will strengthen the goal by clarifying how it will be accomplished. Several examples of these types of "fuzzy goals" are included in the discussion of the sample challenge list.

Assignable harkens back to the prioritization technique of asking, "Is this challenge in our territory?" Does your team have the needed authority to make changes in this area? If the answer is no, go back to the drawing board. There may be cases, however, in which you may not have authority but you do have advocacy for a particular goal—for example, an infra-structure improvement like email to the branches.

In these cases, the goal may be needed to support one of your initiatives, but the actual execution of the goal lies within another group's territory. This is a tricky situation and one that requires consensual role definitions and a close working relationship. If you have a goal like this, the other group must be brought into the process sooner rather than later. No one does well on the execution phase when a task is simply dumped in one's lap.

On the other hand, if the goal is within your territory, then you must determine to whom the goal will be assigned. Assignable also means that you have a group of people for whom this task makes sense, or perhaps a group with the prerequisite skills needed to accomplish the goal. If there can be no clear accountability for a goal, its chances of completion are close to zero.

If there is a group of people to whom the goal can be assigned, but they have not been a part of your planning process up to this point, there is another problem that needs to be fixed. Get them involved immediately; and do not be impatient when they need to go back over ground that has already been covered by the team you originally assembled. Any new people that are brought into your planning process will have their own ideas about what they should do and how to do it—if they did not, they probably would not be in your organization!

Visibility has to do with communicating the goal's tracking information. "Visible" implies that the progress toward the goal should be apparent, like the United Way thermometer, to all those concerned with its accomplishment, not just those responsible for implementing the goal. The measurement or tracking methodology should be understandable, even self-explanatory if possible, and a means of reporting the information should be devised: posted pie charts, email distributions, weekly updates, or whatever makes the most sense for the audience. If it is bad news, better to see it sooner than later—maybe some pressure can be put in the right places to get the progress toward the goal back on target. If it is good news, show it off: it will help to keep the positive momentum going.

Flexibility simply refers to the adaptability of a goal. How easily could it be changed or shifted? Everyone knows how fast-paced the business environment is these days. Changes in technology are revolutionizing both the tools and the structure of businesses. IBM, MIT, UC Berkeley, and Oxford University's recent announcement that they have succeeded in building the first working computers based on a processor consisting of hydrogen and chlorine atoms is an example of the kind of technological revolution that can blow old goals and strategies out of the water.[3]

An organization that has invested in goals that move through the waters of the corporate world like oil tankers will have a difficult time changing direction. It might be better to have a fleet of smaller "goal boats," maybe even gondolas, piloted by individuals who can shift and turn and pull together like a flock of birds. One aspect of goal flexibility might be allowing for diverse methods of accomplishment by smaller goal-directed teams, like Al West's "fluid leadership" concept. This kind of implementation adaptability creates the possibility of realigning a goal, or group of goals, if necessary.

During the discussion of the monitoring phase of our planning process in Chapter 8, the Information Möbius and how feedback assists in the process of keeping goals on target will be discussed. However, any feedback technique will be useless if a goal itself is not elastic enough to be changed.

Inspirational is fairly self-explanatory but difficult to embody in a goal. In the words of Bryan W. Barry, "A strategic plan has little power if it is disconnected from people's hopes and commitments."[4]

An inspirational goal engages the whole self. A previous chapter discussed the whole self and the power that can come from engaging all aspects of yourself and your team members at work. When talking about inspiration, one is really talking about commitment to and involvement in the mission of your organization. Are the members of your team aligned behind the values of your business? Or, if the corporate mission seems too distant, are your team members committed to one another and the goals of your team?

A goal that inspires stirs its implementors to action because they believe in it. It connects people to their hopes and commitments and, therefore, calls up in them their best selves. We all work best when we are doing something we like to do, which does not mean that there are not aspects of any job that are less engaging or less enjoyable than others; it simply means that our jobs make most sense to us when we believe what we are doing has value.

Not everyone can save rain forests, or protect innocent people from violent crime, or rid the world of injustice. And, face it, most corporate goals are about the bottom line—what will the return on investment be? But more and more, businesses are realizing that to mobilize energy, engaging the whole self can be a powerful tool, both for individuals and the corporation.

So, inspiration is really an agreement about the value or worthiness of a goal, and, in its strongest application, there is an almost spiritual component to it. At its root, inspiration is about breath, the heart, feelings; it is not an intellectual understanding about the need for something so much as a physical feeling about the rightness of doing something.

Can your team get behind the mission and values of your company, and does the goal you have identified and the challenge behind this goal embody those values? If yes, then the goal is inspirational.

This characteristic is left for last because it is probably the most controversial and also may not be required in all situations; it is a nice-to-have, not a must-have. But it is a nice-to-have that, when present, can change the action behind a goal from adequate to miraculous.

There is one major exception, however. If your company's mission statement or your goal includes words like "*superior* customer service," or "*exceptional* and *immediate* response to..." or "*driven* by an *obsession* for..." or "*strive* to *delight* our..."—those are the kinds of statements that cannot be accomplished without inspiration. No one puts forth that kind of effort without being inspired to do so. Mission and value statements with these words that are not supported by a corresponding culture that inspires are shams.

EXERCISE 9: TRANSFORMING CHALLENGES INTO GOALS

One note about the process of transforming challenges into goals: taking another look at your team's challenges with the intention of turning them into goals is a preparation for honing in closer and closer to devising the strategic actions that will most effectively reshape the future. Often, the goal can be roughly set, and then be more exactly crafted after the supporting strategic actions are outlined. Sometimes a goal cannot be set without knowing the strategic actions that will be associated with it.

The main point of Exercise 9 is to be as specific as possible in establishing a quantifiable target or goal for the main or root challenge in each cluster. As can be seen in the samples that follow, supporting goals will sometimes suggest themselves in the discussions, and it is a good idea to capture them. These supporting goals can lead directly into or may actually become supporting strategic actions.

At this point, you may want to be time efficient by breaking your team into groups to conduct a first pass of the goal-setting step. If you have grouped your goals by project team, divide the larger planning team into those smaller working groups. If you have grouped your challenges by functionality, a more random group division will work. Put about five to six members in one group and, ideally, no more than eight. Then conduct Exercise 9: Transforming Challenges into Goals.

EXERCISE 9: TRANSFORMING CHALLENGES INTO GOALS

Materials: flip charts, markers, sorted Challenge Lists by clusters.

Duration: 60–75 minutes.

Objective: to transform challenges into effective, quantifiable goals.

Discuss with your planning team the nine characteristics of effective goals. Have a separate sheet prepared which lists the nine goal-aspects and a brief description for each. After discussing them, post this list somewhere in your planning room. This could be a hand-out if small groups will need to work in other rooms separate from one another.

Next, explain the specific assignment: each group will be responsible for discussing the challenges on their list and for restating them as goals that comply with the effective goal characteristics. These newly formed goals should be rewritten on a flip chart sheet and presented to the group for discussion. If the method for measuring each goal is not self-evident, that should also be explained.

Then figure the number needed in each group and break out into groups. Make room assignments, hand out the appropriate list to each group, and give everyone a time to return to the large group room. Depending on the number of challenges each group is handling, I suggest between 20 and 30 minutes for small group work time.

Facilitation tip: Suggest that each group appoint a discussion facilitator, a scribe, and a presenter.

Check on the progress of the groups by walking around and listening in on their discussions. Or, if you have an assistant, split up and try to pass by every working group. Sometimes, groups will have questions about a goal aspect or how to apply it. If necessary, remind the groups about the deadline for return to the large group room.

When the groups return, each group will present its challenges-transformed-to-goals to the larger group. Count on between 10 to 15 minutes per small group presentation, including questions and discussion. As facilitator, gently point out instances in which you think a goal may have failed to meet one of the appropriate characteristics. After each small group makes its presentation, post the finished goal page on the wall.

After the discussions, someone should be made responsible for retyping and distributing the goals and attached challenges. These will be needed for the next step—strategic action formulation.

SAMPLE DISCUSSION FOR THE GOAL FORMULATION STEP AND SAMPLE PLANNING ITEMS: CHALLENGES AND GOALS

As an example of this discussion process, take another look at the final sample challenges carried over from Chapter 6. Some potential points of discussion for each challenge are offered, followed by samples of the transformation of these challenges into goals.

Discussion for Challenge A

> **A. Overall external business challenge: increase our market share in our three-state region by 8% in 2 years.**

Supporting Marketing Challenges

1. Need competitive/customer info: why are customers leaving for Competitor X? Do not know what our current customers like or dislike about our products and services.
2. Do we need new product offerings (i.e., to take advantage of the aging Baby Boomer market)?
3. Not taking advantage of potential new target audience within region.
4. Current marketing collateral needs new look—budget?
5. Design new direct mail campaign? Other distribution vehicles?

This goal looks as if it has a quantifiable portion, but the question is, what will actually be measured? Does the "8%" refer to an overall increase in total market share; or only an 8% increase over the current share of the market. This difference is significant.

For example, say the total market for the products concerned is $100 million, and the sample company commands a 20% market share, or $20 million. Will the 8% increase be applied to the total market, requiring an increase of $8 million to a total market share of $28; or an 8% increase of $20 or $1.6 million, for a total increase to $21.6 million?

Another consideration is whether the market itself is increasing, decreasing, or remaining steady. Are these figures assuming a stagnant or mature market, or is the market growing such that an increase in market share of 8% might reflect an actual *decrease* in relative market share?

Once the measurement methodology for the goal is clarified, in what incremental time periods will the increase be tracked? Will it be spread over the 2-year period by quarter, or simply measured as a year-end goal? Is there a mechanism in place for capturing and reporting progress on this goal? These monitoring issues are explained further in Chapters 9 and 10.

Specifying the "three-state region" seems to indicate that no new market regions will be opened, but that growth goals will need to be met by competing in existing markets.

It appears that challenges numbered 1 through 3 are related to customer response and market data. They will require either additional market research or a compilation and analysis of existing data. If no market research exists, a customer survey vehicle will need to be designed and administered. Thus, the actual goal should reflect the need for research, specify what data are needed, and outline a deadline for obtaining the data.

Challenges 4 and 5 are ideas to look at in considering strategic action—these have to do with possible ways to accomplish the overall goal of gaining market share. They do not need to be their own goals per se; they may or may not become strategic actions, depending on the results of the market data analysis and the strategic thinking that will go on in step 3. So, for now, leave them as challenges until the discussion about strategy in step 3.

Note also that the electronic media—Web site and email challenges in the communications cluster—might be another distribution vehicle or marketing-related program to consider during the strategic planning in support of the overall marketing goal.

Sample Planning Items: Final Challenge A and Goals

After implementing changes indicated by the discussion above, the sample challenges transformed into goals might look something like this. Because of the nature of stating a goal, some of these goals indicate actions that will be reviewed as part of the next planning step. Also note that, as far as it can be known, the person responsible for either completing the goal or supervising its completion is indicated by the initials R.P. (for responsible person).

> ### A. Overall external challenge/goal: increase our <u>total market share</u> by 8% in 2 years, to be tracked quarterly, by increasing sales in our existing three-state region. (R.P.)

Supporting Challenges/Goals

1. Review existing marketing data on the following items by 1Q 20XX:
   a. Customer preferences for current products: sales profile.
   b. Customer suggestions for new products (i.e., to take advantage of the aging Baby Boomer market?).
   c. Reasons for customers' choosing Competitor X's product.
2. If existing data is insufficient, design survey vehicle and implementation plan to capture needed marketing data by 2Q 20XX. (R.P.)
3. Research the feasibility of implementing customer surveys for 5% of current customer base on a quarterly basis, with a go/no-go answer by 3Q 20XX. (R.P.)
4. Design and implement marketing programs or new collateral material in support of overall market share; goal TBD.

Discussion for Challenge B

> ### B. Internal business challenge: improve our communications with other, linked departments, i.e., sales, customer service, fulfillment.

Supporting Communications Challenges

1. Sales department/sales reps too removed from marketing information.
2. Customer service reps overwhelmed by a rash of "new customer" problems during our campaigns.
3. Do we need an intranet Web site for internal customer database and sales info? Or extranet Web site as new distribution channel? (Marketing tie-in above?)
4. Slow or no email possible between headquarters and branches.

Item "B. Improve our communications" is an overall challenge that cannot be measured as it stands. Improving communications is too general and needs to be defined. Fortunately, the following challenges indicate to some degree what improved communications might look like.

Remember from the first derivation of this sample challenge that the problems to be fixed were:

- √√√√ Bad communications with sales department/sales reps: they always get the offer wrong, not prepared for product changes, etc.
- √ Lack of follow-through with our customer service reps, i.e., bad press.

One way to attach a measurement control to this goal might be to tie it to the sales reps' knowledge of new product offers. Is there some way to measure or test their knowledge? A ghost shopper? A formal quiz that each rep must take online before becoming 'certified' to sell? And where should the responsibility for the knowledge transfer reside? With each sales rep individually? With the sales managers for their groups of sales reps? With the marketing department? These are all questions that would need to be answered before the goal can be quantified to target the desired result. The bottomline, however, is that the company's sales performance should improve.

Maybe the goal should be to bring together all members of the fulfillment team—customer service, marketing, sales, and fulfillment—to discuss ways to ensure that the details of a new sales campaign are delivered efficiently and accurately to customers. No doubt the people involved in delivering the message will have some suggestions about why the current system is not working as well as it should be.

The last two items—a better Web site and enhancement of headquarters-to-branch email—will need to be discussed with the IT (Information Technology) team or whoever would actually be conducting the implementation. The goal should be to establish a team to discuss the marketing potential for the extranet Web site. Email is a communications infrastructure issue and will need cross-team discussion and implementation support.

In both of these last items—improvement of the sales efforts and the Web site/email goals—we have started to slide into talking about strategies. Sometimes, assigning a quantifiable goal does require understanding how the goal will be accomplished so that its accomplishment can be measured. This is a good example of how the boundaries between planning steps are in some cases artificial and a reminder that all the planning steps are linked. In fact, some goals can lead into strategies, and some strategic actions will require the revision or restating of certain goals.

Sample Planning Items: Final Challenge B and Goals

After implementing changes indicated by the discussion above, the sample challenges transformed into goals might look something like this:

B. Internal challenge/goal: increase the success of our sales campaigns, measuring the rate of sales conversions of responders (using 1Q 1997 campaign as a baseline) with specific goal TBD based on strategic action program development. (R.Ps.)

Supporting Challenges/Goals

1. Establish a marketing and sales support team by 1Q 20XX—including sales, customer service, fulfillment—with the following objectives:
   a. Review current hand-off practices after new sales campaign is launched.
   b. Consideration of special program for new customers: welcome call and survey, and periodic check-in mailers.
   c. Design and delivery of training for internal personnel who handle new accounts.
2. Meet with IT team to discuss the marketing and distribution potential for extranet Web site and for the enhancement of headquarters-to-branch email capacity. Jan 15, 20XX. (R.Ps.)

Discussion for Challenge C

C. Department challenge: department communications inefficient (too paper-based) and contracting not expedient (takes too much time).

Supporting Administration Challenges

1. Contracting procedure too cumbersome—check with legal?
2. Too much paper in internal communications process? Or, how do we want to communicate with one another: paper, electronic/digital media, face-to-face?

This issue of too much paper is a common one and may not be as easy as it should be to change. Often, excess paper-based communications are a direct result of the culture, especially if there is a "cover-your-ass-and-send-a-memo-to-the-world" approach to information. But sometimes there are approaches that can be taken even within a given culture to mitigate this problem. Some of these are discussed when we follow these challenge-goals into the next step of strategic action.

Note, too, that in the second supporting challenge, there is a hint that what really needs to happen is a discussion about how the department wants to communicate. Maybe paper-based communication is *not* perceived by everyone as being inefficient, but is simply a reflection of how team members like to do business. This idea sounds like it will need further discussion.

But for now, one recommendation for quantifying this kind of goal might be to track either paper consumption or copier usage in the department. Both of these would reflect a reduction of paper-based communications and would be fairly easy to monitor because there are probably already systems in place to track these figures. Another possibility, but more time-consuming since it would require establishing a new system, would be to actually count the average number of items in department in-boxes. Maybe it could be done with a baseline average taken one day a week for four weeks. Another: the same four-day average could be taken after programs for change have been implemented, and the two figures compared.

As for the contracting process: with cooperation from your legal department or whoever handles contracting, you might be able to establish a baseline measurement for either amount of time, number of forms, or number of steps involved. For example, if you determine that a new contract process now takes seven weeks, you could use that data to establish what a reasonable goal would be for reducing that timeframe. You might be able to establish a system for different types of contracts and contractors, depending on the complexity of what is being contracted for. At any rate, a review of the process and establishing some baseline will be critical to judging whether you have actually been successful in streamlining the process. Sometimes, just establishing a means of measurement in itself starts a profitable discussion about the process.

Sample Planning Items: Final Challenge C and Goals

After implementing changes indicated by the discussion above, our sample challenges transformed into goals might look something like this:

> **C. Department challenge/goal: reduce paper use in the department by 15% and copier usage by 10% by 3Q 20XX (R.P.) and expedite contracting procedures (with goal TBD).**

Supporting Challenges/Goals

1. In collaboration with legal, review contracting procedures in order to reduce the amount of time for new contract account set-up by end-of-year 20XX. (R.Ps.)
2. Open broader discussion with department members about preferred communications modes: paper, electronic/digital media, face-to-face, etc. (R.P.)

SUMMARY

As you can see in some of the sample discussions, and as you will no doubt see when you conduct this process with your planning team, it is difficult and somewhat contrived to keep these steps—challenges → goals → strategic actions—separate from each other. Discussion ranges and slips from one category to another. In some cases, specific goals cannot be established until the actual strategic action programs have been developed. In some cases, the challenge itself can serve as a place-holder goal until the implementation details are outlined: in these cases, a goal will need to be established after the strategic action is devised. In other cases, the goal actually suggests the kind of strategic action that will be the necessary first, although perhaps not last, step in the accomplishing the needed change.

Keep in mind that the main purposes of this goal identification step are to engage your planning team in a discussion about what makes for a doable and effective goal, and to help them reflect and focus on the need to establish clear goals as a part of the planning process.

One thing to avoid at this, or any, point in the process, is to get into an argument about the specific wording of a goal or challenge. That kind of wordsmithing can happen offline; it is sometimes more efficient for one or two people to wordsmith independently and then come together to go over their work. Understanding concepts and root problems, discussing issues, and coming to a common understanding of what the task is—these things can only be handled in group meetings. The details of writing—whether it is a document or a series of goals—are best done by individuals and then reviewed by others.

At this point, your planning team should have a list of grouped challenges associated with one or more quantifiable goals. In the next, and pivotal, step—the creation of strategic actions—you will need to roll up your sleeves, engage your imagination, and begin to put together the programs that you hope will solve or ameliorate what needs fixing.

NOTES

1. Quote taken from *The Crossing* by Cormac McCarthy, Random House, New York, 1995, 9.
2. Information taken from the *American Heritage Dictionary*, CD version, produced by WordStar International, licensed from Houghton Mifflin Company, copyright 1993.
3. John Markoff, *New York Times*, Tuesday, April 28, 1998.
4. Barry, Bryan W., *Strategic Planning Workbook for Nonprofit Organizations*, Amherst H. Wilder Foundation, St. Paul, MN, 1986, 19.

8

DEVISING
STRATEGIC ACTION

...there is no superior approach to superb strategy identification than a brilliant intuitive mind.

George Steiner[1]

PREPARATION

If one takes Steiner's quote to heart, then in this step of the planning process, your task as team leader will be to bring out the brilliance, intuition, and creativity of all members of your team. This could mean that guiding your team through the formulation of strategic actions will be a little like herding cats. But take heart; if creativity has a healthy element of chaos, it is also a tremendous amount of fun.

This discussion begins with a few introductory ideas about creativity and how it functions as part of a strategic planning process.

Most people start life as a bundle of imaginative energy. Since one does not know anything, everything is a mystery and every place is a playground. There are no preconceptions about how things should be done, and one uses one's meager skills to explore and solve problems in any way one can. Curious about the nature of something? Put it in your mouth and taste it. See something new, but don't have a word for it? Make one up.

Unfortunately, much of this uninhibited energy gets wrung from us in the process of being socialized at school. One must sit quietly at a desk until a bell rings, follow rules, and speak only after raising one's hand. Along with reading, writing, and 'rithmetic, most schools teach conformity.

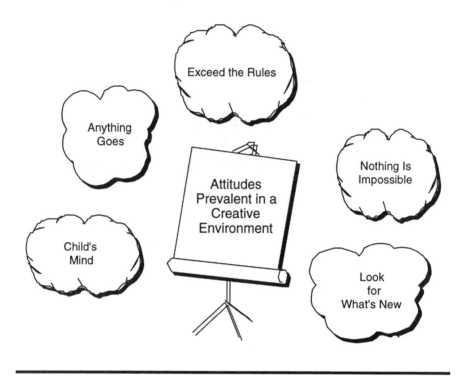

Figure 8.1 Attitudes Prevalent in a Creative Environment

Of course, every environment in which many diverse people come together for a particular purpose must have some rules to order behavior and keep everyone safe. But the trick is to be able to call up one's creative energy in an appropriate way when it is needed—and it is certainly needed for strategic planning.

Proposed here are a few simple rules meant to encourage creativity and intuitive energy in your planning sessions. Figure 8.1: Attitudes Prevalent in a Creative Environment summarizes these suggestions.

Anything Goes

Suspend judgment for at least the initial strategic action brainstorming sessions. This will help create an environment that allows those potentially stifled creative impulses to find voice.

One idea used by Sue Walden and her partner, Carol Rehbock, in their consulting business, Corporate Vitality, is the "Yes and" technique.[2] It is a way to encourage expansive, open-minded discussion and to escalate the production of creative ideas.

An example of how it might work in a conversational meeting setting follows:

> "We could try posting someone to talk to our customers just at the door of the bank on their way out..."

> "*Yes and* we could request that they leave a little taped message to us on a recorder placed in a private area about anything they thought didn't go the way they wanted during their visit..."

> "*Yes and,* we could listen to the messages at the end of the day in our wrap-up meeting..."

In other words, do not veto any idea; add to it. At the beginning of the strategic planning session, take everything into consideration. There will be plenty of time later to use a more rational selectivity and weed out the ideas that are just not feasible. But at this stage, do not squelch anyone or any idea. Say yes to everything: encourage the flow of ideas by building on it and adding one of your own.

It is a well-known fact that, in a discussion, if one person stops talking and the other begins with "But,..." the first person has already turned off and started his or her rebuttal. And it often means that the person starting with "But" has not heard what has just been said. Try "and" instead. This technique also encourages better listening.

Remember: provide a discussion environment that supports an "anything goes" attitude.

Nothing Is Impossible

Cultivate an attitude that questions the status quo. If your team approaches your challenges with the pessimistic attitude that nothing can be changed, there will not be much room for solving problems in a creative way. On the other hand, a wide-eyed optimist who thinks everything is great and is always satisfied will not be motivated to look for a better way to do something.

A better attitude is one in the middle. Stanley Mason puts it this way: "Inventors...are always skeptical but rather than let that skepticism turn into pessimism, we immediately look for a better way of doing something."[3]

Consider the impossible. Most inventions, particularly those that most radically changed the world, were revolutionary because they seemed unthinkable at the time. Flying? A computing machine? Cloning? Men on the moon? Impossible!

Exceed the Rules

Again, people have learned over a lifetime of following orders and suppressing their own feelings that they must follow the rules set by the powers that be. By "exceeding the rules," I mean explore why the rule is in place and understand it enough to expand it, or recast it in a way that retains the truth of its necessity.

Perhaps another way of saying this is, "Some rules can and should be broken." There is a difference between understanding the basis for and actively respecting authority vs. passively acquiescing out of fear or simple disinterest.

Always question authority and make it a habit to ask yourself, "What is the truth in this situation or information?" This requires personal integrity and emotional detachment that is embodied in the concept of CHI, as discussed in Chapter 3. This attitude also requires boldness and a willingness to step out of line when things are not going the way you feel they should.

Look for What Is New

Always be on the lookout for what's new: new experiences, tastes, sounds, ideas. Be observant of the details. Exercise your mind to accept or consider accepting things that are new or different from the ordinary.

Maybe one way of instituting this idea in the strategic planning session is to meet in a new or unusual place, or to ask different sets of team members to work together. You might pose the question—"What would make our strategic planning session creative?"—to the team and see what they suggest might offer them new experiences and set the tone for a quantum leap in creativity during this strategic action planning step.

Maybe you should all go to the zoo for a picnic during the day and work on your planning session in the evening. Maybe the session should be at your local science museum after-hours. Maybe you should convene at the airport. Or a winery. Perhaps the session should start with a meal from another culture that is unusual and unfamiliar to most team members.

Try something new at your meeting as a metaphor for creating and using this attitude in your planning session.

Child's Mind

One technique that Stanley Mason recommends for sparking creativity is the idea of putting one's mind inside other things. He says, "How would it feel to be inside a lawn mower when the starter rope is pulled?"[4]

As you might imagine, a child will have a ready answer for such a question. A child approaches tasks with innocent playfulness and few judgments. He or she simply explores and experiments freely. It is this aspect of a "child's mind" that you want team members to bring to the strategic action discussion.

Talk some of these ideas over with your team before you begin the strategic action planning, or devise your own set of principles for encouraging an atmosphere of creativity and insight.

If your team members agree to a set of guiding principles in your strategic discussion, post these on the walls of your meeting place. They can be used as reminders if nay-saying begins to erode the expansiveness of your session, and they might also spark some strategic ideas of their own.

EXERCISE 10: INFLUENCES REPRISE

In Chapter 3: Becoming a Strategist, Exercise 3 (called Influences from the Future) prompted you to think about future considerations that could have a bearing on your current planning project. In Chapter 2: Beginning Concepts, a similar exercise (Exercise 2: Influences from the Past) asked you to jot down influences from the past that could affect your project.

Take a look at both of your lists again. There may be elements on these past and future influences lists that have already been incorporated into the challenges and preliminary goals that your team has devised. However, if there are items that are not reflected there, compile a list of these to discuss with your team. Use this discussion as an opportunity to add to the list any other influences, trends, tendencies, problems, or opportunities that you want to take into consideration as you begin to devise your strategic actions.

This is really a recapitulation of the "situational analysis" step that will give your team a chance to pull together ideas or influences that determine the parameters within which your strategic actions must be created.

If you feel strongly that you want your team to have the full benefit of producing their own influences lists and/or you have the time to do so, go back to those chapters cited above and conduct the exercises with your entire team. If this is not feasible, a group discussion, based on your influences list, should suffice.

EXERCISE 10: INFLUENCES REPRISE

Materials: raw lists from your earlier Exercises 2 and 3; flip chart and markers.

Duration: 45 minutes.

Objective: to discuss with your planning team influences from the past or future that could affect your project.

Either conduct Exercises 2 and 3 with your group, or engage them in conversation about past and future influences that could affect your planning project. As discussed before, these influences could be resource issues, business or industry uncertainties, personnel problems, or (if part of this exercise is private) even personal issues.

Ask team members to draft their own lists individually, then ask for a group sharing. Make a list on flip chart pages of what your team feels could influence the outcome of your planning project—either positively or negatively.

After completing this influences discussion, post the final compiled list on the walls in the place where you will be conducting your strategic action sessions. Having these elements visible may spark some ideas that could otherwise be missed.

DEVISING STRATEGIC ACTION

You may want to review the concept of strategic action with your team. A more thorough discussion of its aspects was developed in Chapter 2: Beginning Concepts. In a nutshell, strategy is combined with action in the phrase "strategic action" to indicate that this activity is selected for its direct effect on your goal/challenge. A strategic action is like an arrow shooting for a target—it has one objective in mind.

One of the linguistic roots of strategy is the Latin word *agere*, which actually means to do or act. This root word is a perfect metaphor for the concept of strategic action because it emphasizes that the strategy carries the action within it. That is why in the steps of our planning process, strategy is not separated from action—they are combined into one powerful dynamic: action with insight.

So, to begin the strategic action brainstorming, first make sure that you have the right group of people present. You may need to add to your core planning team at this point because, ideally, your planning group should include everyone on the strategic action phase who will be assigned a specific challenge or goal to implement. If not all of the execution team can be present, then try to get department or division representatives or key influencers.

For the following exercise, divide team members into groups of five or six. If possible, try to have more than one group brainstorming on each challenge/goal. This allows for a variety of ideas that can often be combined to create a truly winning strategic approach.

There are two basic tasks that must be completed in this strategic action step of the planning process:

- Outlining a broad-brush strategic approach for each challenge/goal
- Devising individual strategic actions that implement that strategic approach

A discussion of each of these component concepts follows.

Outlining Your Strategic Approach

In order to outline your strategic approach, your team will devise what is called "Guiding Assumptions." These are the principles, beliefs, or conditions underlying your particular business or organizational environment. These assumptions define a system and determine its behavior; they are the parameters that limit and direct your team's set of strategic actions.

Sometimes, these assumptions have to do with trends or developments in your industry. For example,

> The next technological revolution in chip manufacturing won't begin until possibly 20XX at the earliest, 20XX at the latest. Next generation chips will begin to appear in the marketplace by ?Q 20XX.

Another category of assumptions might be reflective of the general economy or economic indicators within your market or industry:

> We are assuming that growth in our product sector will flatten by the end of 20XX and remain stagnant into the next decade.

Sometimes, the guiding assumptions can be related to your organization's core competencies:

> Our strength in the current market is our technological advantage and our strong R & D department.

These are the background conditions that will provide the structural underpinning for your strategic actions. Some of these ideas may already have been discussed or alluded to when your team analyzed the world of challenges, beginning with your grassroots situational analysis. Or, they may be conditions or influences that have been included in your Influences Reprise exercise (Exercise 10). Often these assumptions are a kind of summing up or shorthand for the detailed and particular aspects of the economy, your industry, and your organization within that industry.

These assumptions are often provided to your planning team by your management leaders, or they are outlined for you in higher-level strategic planning documents. If that is the case, it is still a good idea to repeat them or restate them before you begin this next strategic planning exercise.

One of the critical reasons for considering and identifying these underlying assumptions is that they provide triggers for changing direction. If business analysis or simply the passing of time indicates that a stated assumption proves to be wrong, the implementation plan and the activities supporting that plan will have to be adjusted accordingly.

Strategic Actions and Tasks

Once your implementation team members have established their list of underlying or guiding assumptions, the brainstorming on strategic actions can begin in earnest.

This is the point in the planning process when your team will begin discussing activities and details—the "HOW" of your plan:

- What will we do?
- How will we do it?
- When will we do it?
- Who will do it?

These are the questions that will be answered during the creation of your strategic actions.

EXERCISE 11: INTENTIONAL INSIGHT

The following exercise is called *"Intentional* Insight" because, in it, the use of insight is directed specifically to your planning task. Some creative thinkers will tell you that their best ideas come to them in the shower or while they are taking a walk. But there is a place for will or intention in the process of creating. You and your team need to get a job done; so you must call up your powers of creativity now.

Insight has been discussed in a general way—how one might nurture and use it, what environments might encourage it—but, as indicated by the quote from George Steiner at the beginning of this chapter, in essence, this is where the miracle happens. No one can codify an approach that delivers a sure-fire formula for innovation and strategic brilliance. Some entrepreneurs are gifted with a kind of sixth sense—a high level of ability and success in understanding the direction of the market or a way of perceiving influences that gives them the foresight to make the right business moves. More than likely, your planning teams will include some individuals who excel at strategic thinking.

The following exercise provides some guidance in supporting strategic thinking and allows everyone to bring their experiences to bear in formulating strategic actions.

EXERCISE 11: INTENTIONAL INSIGHT

Materials needed: enough copies of the challenge/goal clusters to have one for each team member; flip charts, markers, tape.

Duration: approximately 2 hours.

Objective: to establish guiding assumptions for each challenge/goal and to devise individual strategic actions or tasks that implement that strategic approach.

Begin this exercise by establishing the principles for these creative brainstorming sessions. These attitudes or principles could be the ones discussed previously in this chapter in the 'Preparation' section, or any that the group itself chooses in order to encourage creativity and insight (10–15 minutes).

Outline and discuss the objectives of this meeting and the specific outcomes that will be required from each small group (15–25 minutes).

Facilitation tip: suggest that each group be responsible for reporting back to the large team meeting on two strategic actions: (1) their most outrageous or outlandish idea, and (2) the one that they think is the most doable. This may encourage a spirit of creativity and fun.

Go over the timeline: each group will have from 30 to 45 minutes to outline their actions. They can go into as much detail as they have time for in terms of implementation, budget, and timelines. Again, suggest that each group appoint a facilitator and a scribe.

Send the groups off to brainstorm.

Check on the groups as time passes to make sure that the timeframe is right. If everyone seems to think more time is needed, establish a revised "return to group" time and communicate it to all working groups.

In the large session, allow each group between 5–8 minutes to present their two approaches to the assigned goal. Try to give groups working on the same challenge/goal consecutive presentation slots.

Gather the information from all groups and appoint a smaller group to sort and process the information into a standard format in preparation for the next group exercise.

EXERCISE 12: ON-TARGET ACTION

This final strategic action step serves as a checkpoint to ensure that actions are on target against challenge/goals. This is a chance in full group discussion to review the results of Exercise 11 in order to iron out any inconsistencies, verify a common understanding of goals and solutions, make sure that efforts will be expended to accomplish the right goals, and identify preliminary areas of strategic action crossover between teams or departments.

The results of this exercise could become the first draft of a planning document. Chapters 9 and 10 will further discuss the document itself, both as a communications vehicle and as a monitoring and tracking device.

For now, approach this final step in the strategic action phase as a last-chance checkpoint and a way to verify group alignment on strategic actions.

EXERCISE 12: ON-TARGET ACTION

Materials: flip chart and markers, draft of strategic action ideas from the previous exercise.

Duration: 60–75 minutes.

Objective:

1. Review assumptions and strategic actions for direct applicability to goal.
2. Review phraseology and clarity of tasks outlined.
3. Identify any issues or dependencies that cross implementation team lines.
4. Discuss what modifications or measures could be taken if guiding assumptions change.
5. Outline ballpark budget figures and timelines for completion of actions. (This could also be completed offline by a subgroup.)
6. Appoint one member of implementation team to be part of larger project monitoring team (further discussion in Chapter 9).

Bring together, if possible, all implementation team members for each challenge/goal cluster. Hand them the ideas drafted from the last session and assign the tasks outlined above.

Regarding the appointment of an implementation team representative, discuss the importance of this project monitoring group to the success of the entire strategic planning process. The composition of this group will be critical.

Give small groups between 40–60 minutes to accomplish their tasks (depending on the complexity of their action lists). There will be no reporting back on specific small-group decisions, but any discrepancies or implementation issues that cross implementation group lines should be raised in the larger group and recorded.

As always, assign one or two individuals to gather revisions and changes to strategic actions. These team members should reproduce the results of this meeting for return to all team members.

FINAL CHECK MOP-UP

Even after Exercise 12 is completed, depending on the level of detail needed, your strategic actions may still need to be handed off to other teams for supporting task and subtask development.

There are many action steps that go into the achievement of a broad-based corporate goal. Each of these tasks or task groupings should have its own accountable person, measurements, and completion deadline. Also critical is drawing up an interdependencies or linked-task list for those tasks that require implementation across team lines or hand-off midway through completion.

These mop-up tasks should be assigned to appropriate teams or team members and completed before the start of Exercise 13 in Chapter 9. This might be a good time to share your preliminary planning team efforts with other persons who have not been part of the core planning team. Clearly mark these ideas as "DRAFT" before you distribute them.

Chapter 9 discusses the Project Implementation Team, whose members should have been designated in Exercise 12. This group's task will be to shepherd the implementation of the strategic plan through to completion.

SAMPLE PLANNING ITEMS: GOALS AND STRATEGIC ACTIONS

After completion of the above exercises, your challenges and goals might evolve into the following strategic actions. Note that some items previously identified as goals have been wrapped into strategic actions.

A. Increase total market share by 8% in 2 years, to be tracked quarterly, by increasing sales in our existing three-state region by roughly 1% per quarter beginning 2Q 20XX. (R.P.)

(Base measurement: total market share as of year-end 20XX.)

Guiding Assumptions

Our core competencies focus on people skills and strong branding, not technology or low-price leadership. Therefore, growth will be achieved by deepening relationships with current customers and by partnering with them to gather intelligence in order to increase our customer base.

Major thrust will be a renewed customer service orientation in preparation for potential new product design and launch (beginning 4Q 20XX).

No dramatic monetary offers will be utilized at this time, except the perceived value of courtesy and closer personal contact.

Strategic Actions/Tasks

1. Customer intelligence survey team tasks include the following:
   a. Identify survey research and design team by 4Q 20XX. (R.P.)
   b. Review existing marketing data on the following items by 1Q 20XX: (R.P.)
      - Customer preferences for current products: sales profile
      - Customer suggestions for new products
      - Reasons for customers' choosing Competitor X's product
   c. If existing data is insufficient, design survey vehicle and implementation plan to capture needed marketing data by 2Q 20XX. (R.P.)
   d. Assemble production team to research printing and distribution mechanics to have survey in customers' hands by mid-1Q 20XX. (R.P.)
   e. Develop analysis tools to evaluate current customers and understand them using finer demographic customer profiles, based on survey results, 2Q 20XX. (R.Ps.)
   f. Follow up with "We Need You" calls for portion of customers who do not return surveys by end 3Q 20XX. (R.P.)
   g. Research the feasibility of implementing customer surveys for 5% of total current customer base on a quarterly basis. With a go/no-go answer by 3Q 20XX. (R.P.)
   h. Hire mystery shoppers to shop in areas of projected expansion and report back to marketing implementation team by 2Q 20XX. (R.P.)
2. Marketing campaign redesign and execution team tasks include the following:
   a. Develop 3Q marketing campaign with a diverse array of approaches—including initial contact vehicle and series of follow-up tools—to match unique, customer-specific demographic niches. (R.Ps.)
   b. Research the utilization of Web site for partnering program using permission marketing techniques, beginning 2Q 20XX. (R.P.)
   c. Consider new series of testimonial ads in target market area, mid-2Q 20XX. (R.P.)

      d. Pursue partnership with retail outlets to institute "Open House" tables and demonstrations at local outlets with the purpose of establishing personal contact with customer base, mid-3Q of end 4Q 20XX. (R.P.)

      e. Consider the formation of "You Tell Us How" teams of consumers to advise marketing group, in exchange for special customer offers, "Thank You" gatherings, and appreciation awards, 2Q 20XX. (R.P.)

  3. Training design and delivery team tasks include the following:

      a. Co-design, with sales, generalized training that reflects new tactics and attitudes about customers, 1Q 20XX. (R.P.)

      b. Work closely with marketing campaign team to design suitable training for new programs, goals TBD.

Potential Cross-Team Issues

Continue to nurture closer working relationship between marketing and sales: form joint implementation team?

Need support from training department on customer care training design and delivery.

Need to work with personnel on hiring criteria to reflect new attitudes?

B. Internal challenge: increase the success of our sales campaigns, measured against rate of sales conversions of responders (using 1Q 1998 campaign as a baseline) with specific goal TBD based on campaign targets set by marketing team. (R.Ps.)

Guiding Assumptions

We believe the weaknesses in our current sales campaigns are the result of two general factors: ineffective campaign execution and the need for a revised campaign approach.

Specifically, we want to improve the following: (1) the break-down of internal communications between marketing and sales; (2) inadequate follow-up after the initial customer contact has been made; and (3) a "one-size-fits-all" approach to the customer (note crossover with overall marketing goal A above).

Strategic Actions/Tasks

  1. Marketing/sales joint task force: establish a marketing and sales support team by 1Q 20XX—including sales, customer service, fulfillment—with the following objectives: (R.P.)

    a. Review current hand-off practices for sales campaigns, 1Q 20XX. (R.P.)

    b. Consider and/or design special program for new customers: welcome call and survey, and periodic check-in mailers, 3Q 20XX. (R.P.)

    c. Design and delivery of training for internal personnel who handle new accounts, 2Q 20XX. (R.P.)

2. Email tech team tasks: support/initiate project for the enhancement of headquarters-to-branch email capacity. Jan 15, 20XX. (R.Ps.)

3. Web site task force task: assess computer sophistication of current customers and, if warranted, meet with IT team to discuss the marketing and distribution potential for extranet Web site, 2Q 20XX. (R.Ps.)

Potential Cross-Team Issues

IT and Marketing/Sales initiatives will require cross-team support.

IT may also need to be involved in current customer computer-use analysis.

Will there be shared budget issues on potential extranet initiatives?

C. Department challenge: reduce paper use in the department by 15% and copier usage by 10% by 3Q 20XX (R.P.) and expedite new contract account set-up from 8 weeks to 3, by end-of-year 20XX. (R.Ps.)

Guiding Assumptions

The assumption is that increased paper and copier usage can be combatted by environmental education and a focus on environmentally-friendly practices in the department.

One also assumes the Legal Department will work with us to consider and improve the efficacy of contracting procedures.

Strategic Actions/Tasks

1. Contracting procedures task force: in collaboration with Legal, review contracting procedures in order to reduce the amount of time for new contract account set-up from eight weeks to three, by end-of-year 20XX. (R.P.)

2. Reduce paper use task force:
   a. Open broader discussion with department members about preferred communications modes: paper, electronic/digital media, face-to-face, etc. At next department meeting. (R.P.)
   b. Suggest at next quarterly division head meeting that a task force be appointed and a series of environmentally-friendly policies be developed, 1Q 20XX. (R.P.)
   c. Research national or local environmental organizations to see what material already exists, 2Q 20XX. (R.P.)
   d. Research broader applications of an environmental policy: copier cartridges, laser-jet ink replacements, recycling of all office waste materials, etc., 2Q 20XX. (R.P.)
   e. Disseminate information and begin campaign by 3Q 20XX. (Wrap decreased copier and paper usage goals inside the corporate campaign.) (R.P.)

Potential Cross-Team Issues

To be successful, the environmentally-friendly policies must be adopted corporate-wide.

SUMMARY

Devising strategic action is probably the most difficult planning step to systematize. Your team is unique in its make-up of individuals; and the talents and skills of each individual will determine how effective your team's strategic actions will be. Certain synergies will occur during your strategic planning session that will be completely unplanned and unreproducible. And because there is always a little bit of magic involved in crafting effective strategies, the exact progression of this process is difficult to outline or predict.

The bottom line is that, by the end of this penultimate step in the planning process, your team should have generated concrete actions or tasks that address the most important challenges facing your group. In most cases, these goals will also have measurable components that can be tracked as the team progresses toward the goal.

Whether your strategic actions will actually provide the results called for in your goals is yet to be seen. Supporting, monitoring, and measuring that hoped-for success is the subject of Chapter 9.

A few final thoughts on your process. It is likely that every team will have at least one 'doubting Thomas' member. At a certain point in most planning meetings, I am generally that skeptic. I want to say, "Show me—how do I know that will work?" or "What if this doesn't go the way we planned?," or "Now, what are we trying to solve with this activity?"

This is the point in your planning process when you might want to initiate a discussion that features the skeptic. You know the old adage, "Anything that can go wrong, will go wrong." So what will those missteps be? Take some time with your group to brainstorm all the possible ways the plan could fail. What could go wrong?

Some planning teams like to prepare a contingency or backup plan that outlines possible fallback actions if the proposed activity fails. This was suggested as one of the tasks in Exercise 12. Preparing such a plan can be very useful because if things do go wrong, you will not be starting your problem-solving with a blank slate. This kind of skeptics' discussion also provides one last chance to give voice to the nay-sayers on your team.

Once this "what if?" session takes place, however, it is important to regain your positive momentum as a team. In fact, once the plan has been finalized, the main job of all team members will be to remove any and all barriers to its success. When implementation begins, you and your team members will need to be cheerleaders and wise guides for the expanded group of employees who will be asked to participate in the implementation of the planning tasks.

But there is one more step before implementation gets underway. Push the mental pause button; congratulate yourself on getting this far. Then take a breath and go on to Chapter 9: Monitoring and Measurements.

NOTES

1. Steiner, George, *Strategic Planning: What Every Manager Must Know*, The Free Press, Macmillan Publishing Co, New York, 1979, 188.
2. Carol Rehbock and Sue Walden are co-founders of the consulting firm Corporate Vitality. They offer workshops that help build a positive work environment; they can assist with change management, spark your team's creative juices or engender a heightened awareness of customer service. Sue Walden can be reached in San Francisco, California, at 415-885-5678 (voice) or 415-885-4314 (fax).
3. Mason, Stanley, "Lessons from the Inventor," *BottomLine Personal,* June 1, 1998, 9.
4. Mason, Stanley I., *Inventing Small Products*, Crisp Management Library, Menlo Park, CA, 1997, 88.

9

MONITORING AND MEASUREMENTS

The chairman of the American Productivity and Quality Center tells a story about a big-company CEO who, in a moment of contemplation, revealed a deep desire, "I wish we knew what we know."

Justin Hubbard[1]

Imagine you are standing in the middle of a large, square warehouse with blank walls and floor. On one distant wall is a small door—the only exit. Your boss opens the door and calls to you, "Look, I'll meet you in the room next door. OK?"

The door slams shut and just as you begin walking toward it, the lights go out. Surely, there has been some mistake, you think. But there is nothing to do about it now. You proceed carefully, one tentative step at a time, toward that door. But at each step, you wonder—"Have I turned slightly to the left? Is the door still in front of me?"

Because of the dark and the strange feeling of isolation, you have no sense of where you are, where you have been, or where you are going. But you cannot go back and start again. That would just worsen your situation. You move slowly, although you really want to run. Running might get you there more quickly, but where is "there" exactly?

This little story is an analogy for trying to meet a goal in an environment where feedback does not exist. The only information one has the benefit of in this narrative is one's direct experience of what is happening in the moment. Since there is no concrete information available about the past or the future, the only way to solve the problem is simply to persevere and be alert to present information. One must simply proceed and hope for the best, perhaps hitting a wall eventually, then making a decision to traverse it either clockwise or counter-clockwise to find that illusive exit door.

Of course, this is an absurd situation—more like a bad dream than a planning scenario—but it is not far from what can happen in planning environments after an action plan is written and implementation begins.

An enormous amount of energy is expended in creating a plan. Many key people come together for long hours of discussion to wrangle about the details of what to do when. But too often, after the action steps have been devised, everyone thinks the job is done. And, later, there is much consternation about why many of the goals were never met.

The problem is not the plan; the problem is the process. The last, crucial step before beginning implementation—establishing effective monitoring devices—has been omitted.

So, an examination of some methods for making sure that your goals will be achieved is in order.

MONITORING: WHAT AND WHY

What exactly does monitoring mean? What needs to be monitored and why?

Monitoring is the control step of the planning process. It is a process and set of procedures for tracking exactly what progress is being made against a specific, quantifiable point or goal. Monitoring should also include a continuous assessment of the current conditions. Whereas the identification of the challenges, changing them into goals, and devising the strategic actions have all emerged from the theoretical, intellectual processes of your group, the monitoring and measurement against these goals is a reality check. Monitoring is about what is happening "now." This monitoring in and of the "now" allows your planning team to continuously adjust the theoretical in order to take into account information that you did not have at the time your plan was devised.

As an example, suppose you work for a power company and your planning team plotted out a goal and deadline based on an assumption about work load and assigned personnel. Halfway through the implementation of your strategic actions, a huge storm brings down your system and requires special emergency customer service teams to fix the problem; everyone is taken off current projects until all customers are back online. That is a condition, based on the reality of "now," that will require an adjustment in some goal timeframes and perhaps other aspects of the plan as well.

To monitor effectively, one must know what to monitor and how to monitor it; and, most critically, one must be stringently honest: a denial or misinterpretation of what is actually happening can be devastating

The monitoring function is multifaceted. Its purposes include the following:

- Providing communication tools that track the effectiveness of strategic actions toward a goal and, therefore, providing feedback for the implementation team and its greater community
- Providing measurement standards for managerial performance
- Providing guidance for business decision-making
- Triggering directional or strategic action adjustments, if needed, based on an assessment of current conditions and allotted resources

Each of these purposes for monitoring require specialized monitoring devices and involve different considerations.

MONITORING AS A COMMUNICATIONS TOOL

As noted in the story of the United Way goal thermometer, communicating team progress toward a goal can be an effective motivator for success. Heisenberg's uncertainty principle states that simply the act of observing a phenomenon *changes* the phenomenon. Monitoring is observing and, as such, changes what is being monitored simply by its very existence, particularly if the information is communicated broadly.

Communicating progress being made toward a specific goal accelerates the inherent, current results. In other words, if good progress is being made, communicating that progress can have an inspirational and motivating effect. But if a team falls far behind in its expected achievement against a goal, broadcasting those results could be demoralizing and tend to reduce even further the team's efficacy.

A good example of the accelerated positive effects of monitoring comes from a story about Bentley Mills,[2] a rug manufacturer operating in Los Angeles. At Bentley, the amount of carpet scrap had never been measured until June 1996 when the CEO, Ray Anderson, began focusing on sustainable business practices. When the measurements were made, employees "discovered they were throwing away 2 running yards of carpet for every 100 yards they stitched." In some ways, by simply making that measurement and spotlighting their waste, change was catalyzed. By year-end 1997, scrap had been trimmed by 30%.

On the other hand, one can probably cite an instance when a cycle of negativity began and became exacerbated by the communication of negative results. When sales in one of my client's business units started dropping unexpectedly, people were at first surprised but resolved to work harder. But the bad news continued and, after three quarters of falling statistics, two of the lead sales representatives posted to other units; expectations for success within the division dropped; and the unit manager left for another company. Communicating the negative results affected everyone involved. Unfortunately, assessing and understanding the real causes for the drop in sales came many months later.

This is a not an argument for keeping statistics secret. On the contrary, the point is that monitoring business activities and communicating those results have a strong effect on the attitudes of the people involved. Monitoring and creating effective communications vehicles for the monitoring function are powerful tools for change.

If a team is not meeting the expectations it had for its own performance against a specific goal, there are a whole raft of reasons that may need to be explored. Sometimes, it will become apparent that reaching a goal is not possible. The goal itself may be faulty, or perhaps conditions have changed such that a complete revision of the plan is needed. Poor performance may not be a negative factor if it is taken as information that must be examined; it may be simply providing an opportunity to make needed adjustment.

One of the most powerful monitoring communications tools is the planning document itself. Often, a concise and readable compilation of the team's work can contribute to the ultimate success of the plan. The more knowledgeable employees are about the details of the strategic plan, the higher are its chances for success.

Techniques and considerations for determining the most effective communications tools for monitoring feedback are addressed later in this chapter in Exercise 13: How Do You Measure Success?

Some of the questions your team will need to answer are the following:

- What will success look like?
- What kind of measurements will be tracked?
- When will the measurements be taken? How often?
- When will the measurements be reported? How often?
- Who will do the tracking computation, analysis, and reporting?
- Who will receive the tracking information? Who needs to know what?
- How will the needed information get to the right audiences?
- In what format?

MONITORING AS STANDARDS FOR PERFORMANCE

Everyone by now is familiar with MBOs, the acronym for "managing by objectives," a concept attributed to Peter Drucker. What MBO actually means is that everyone in an organization has job objectives that are tied to specific corporate goals in a company's strategic plan and, to be most effective, are tied to that individual's compensation.

Completing one's personally assigned objectives can mean a bonus or promotion, while not making objectives could mean some kind of disciplinary action or loss of compensation.

In a well-structured MBO environment, the corporate goals and the terms of their measurement establish the foundation for the formulation of these employee objectives. It is a kind of trickle-down process. Corporate goals are broken into smaller and smaller tasks or subtasks, each with its own measurements and timelines attached. Each employee at an organization is responsible for his or her appropriate level of achievement of those goals.

Use of a strategic plan for management and job evaluation criteria varies by organization. It is important to understand at the outset how or if these strategic actions and their measurements will be used in evaluative situations with employees. Particularly if the strategic actions are used to create an MBO-style accountability, it is important to involve employees at all levels, as soon as it is feasible, in their part of the strategic plan and its implementation. As previously discussed, buy-in is critical to the success of your planning process and strategic accomplishment.

The types of monitoring techniques for this particular purpose need to be those that report on follow-through, accountability, effective use of resources, and intermediate deadlines.

MONITORING AS GUIDANCE FOR DECISION-MAKING

Probably one of the most important reasons for developing a strategic plan is to capture the essential thinking and positioning for an organization. In this sense, the process of monitoring provides the feedback necessary for effective decision-making.

As Peter Lorange states is his very practical and classic planning text, *Strategic Planning Systems*, "An effective strategic planning system must help line managers make important decisions...managers are not interested in plans; they like making decisions."[3]

Perhaps a truer statement was never spoken. No manager likes to take time away from "real work" to participate in a planning process. But the payoff is that a good plan will make the real work go more smoothly. An effective strategic plan and planning document should assist a manager in making decisions about how best to use both monetary resources and staff time. And in the dog-eat-dog business environment, a solid decision-making framework is indispensable. But the strategic directions outlined in a plan are not as useful to a manager on a day-to-day level as the actual measurements and feedback on specific goals that monitoring tools can provide. An outline of strategic direction is useless in a void. (Recall the "meet me in the next room" scenario at the beginning of the chapter.) Strategic direction coupled with data is what assists a manager in making the right decisions.

Imagine a scenario in which the team working on the new marketing material is behind schedule. The team leader has approached you, the manager, to find out if he/she can hire an extra person to assist with the project. In the back of your mind, you know that the most critical goal for your unit is the completion of the system conversion and the networking of your department's computers with those in the rest of the organization. How do you weigh the needs of your department's compliance on the system conversion with the request for additional resources from the collateral material project team?

An effective strategic plan will provide a framework for discussion of that resource problem. But again, it will be the monitoring against goals that may help you in finding a resolution. You need real information about what action is most effective in the process of working to achieve specific goals.

How far off-schedule is your collateral team? How far along is the department infrastructure group in their networking task? Are the two teams on-budget or not? Is there another project team whose resources you could use temporarily? And what is the relative importance to the organization of the completion of these projects?

Effective monitoring provides a framework for good decisions.

MONITORING AS STRATEGIC ACTION FEEDBACK

Probably the most important function of monitoring is the one that diagnoses the continuing validity of the strategic plan itself. The strategic actions are based on certain assumptions about conditions internal and external to your organization. You and your team members made your best guesses about likely scenarios, trying to build into your plan the most doable and resource-effective actions guided by those business assumptions. This aspect of the monitoring function provides a reality check on those assumptions.

The kinds of questions that monitoring tools for strategic action should be able to answer include the following:

- Are the underlying business assumptions that your strategic actions are based on still valid?
- Is there an industry trend or development that you missed in creating your strategic assumptions?
- Are there new competitors or products in your market niche?
- Is one of the scenarios you thought least likely the one that is coming into being?
- Is the action your team is taking working out the way you thought it would?

As mentioned previously, it may also be the case that projections about resources are now inaccurate. Perhaps your team assumed that a certain number of personnel would be devoted to working on the strategic actions outlined by your planning team, and the reality is that people have been reassigned to other projects. In the context of resource assessment, these are the kinds of questions that your monitoring process should be able to answer:

- Is the deadline for this goal still realistic?
- Are personnel resources sufficient to achieve this goal on schedule?
- Are the people working on this strategic action properly deployed?
- Do we need to revise any goals or deadlines?
- Do we need fewer or more resources—whether personnel or budgetary support—to accomplish this goal on schedule?

In the planning process, your team identified possible back-up activities that could be considered if the selected strategic actions were determined to be unsuccessful. But at what point will that back-up plan be brought into play?

The monitoring tools you devise must be able to give you and your teammates the feedback needed to adjust the plan itself. They should be linked directly to your guiding assumptions such that certain 'flash points' trigger the need for a dramatic shift in direction; or, at the least, a major reconsideration of strategic direction.

Chapter 10: Implementation Tips discusses a process for continuous feedback that will help in catching these shifts or changes in basic strategic assumptions.

HOW DO YOU MEASURE SUCCESS?

So, having discussed in a general way the purposes of monitoring, one can now get down to brass tacks on how to accomplish the functionalities outlined above.

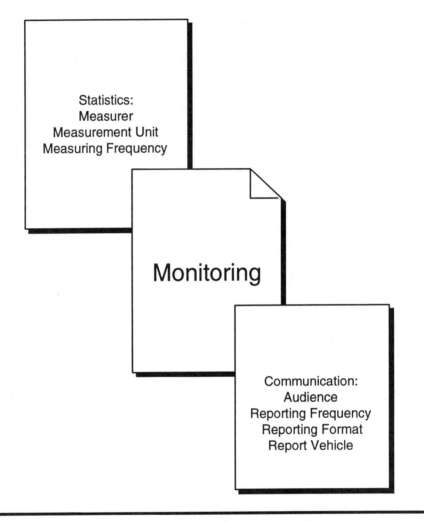

Figure 9.1 Considerations for Creating Effective Monitoring

Figure 9.1 illustrates the various aspects to be considered when evaluating what kind of monitoring tool will be appropriate for a particular goal.

The aspects are divided into two categories; the first category has to do with the actual **statistics**:

- Measurer
- Measurement unit
- Measuring frequency

The second category of monitoring components has to do with **designing the best communication vehicle** for the statistical information. In this second grouping, the following characteristics must be considered:

■ Audience
■ Reporting frequency
■ Reporting format
■ Report vehicle

In the first category, the **measurer** refers to the person who will be responsible for gathering the monitoring information. Make sure someone is responsible both for calculating the tracking number and passing it on to whoever will be producing the monitoring vehicle, if this is a different person.

The **measurement unit** refers to a "precisely specified quantity in terms of which the magnitudes of other quantities of the same kind can be stated."[4] In other words, it refers to the actual method for collecting the datum or data. What will be measured? For instance, one of the goals stated that paper usage would be reduced by 15% by 3Q 20XX. But how will that 15% be measured. Most copiers have a counter, so in this case, the measurement unit would be one copy as counted by the machine's internal mechanism. If the measurement unit is clear, is there a mechanism in place to produce that measurement or will a process need to be created to make the calculation on new bookings percentage growth, number of new accounts, or increase in bookings? In something even as simple as the copier usage, someone will need to read the number from the machine and devise a standard time for doing so—close of business Friday, for example—in order to make the calculation valid.

The **measuring frequency** has to do with the periodicity of the measurement. How often will this number be recorded? Every week? Once a month? One thing to take into consideration in establishing the measuring frequency is the total timeframe for completing the goal. If presently in the 2Q and the goal must be completed by 3Q, a once-a-quarter measurement is probably not going to do the trick. Weekly would be more like it. On the other hand, if the goal is increasing the market share by 8% over the course of 2 years, a weekly measurement is overkill. The measurement unit and frequency should fit the scope of the goal.

In the second category of monitoring aspects—designing the communication vehicle for the data—the audience is the first consideration. **Audience** refers to whoever the monitoring information is directed or will be reported to. Just as in any communications plan, the audience is multifaceted and could include recipients both internal and external to the organization. One question to consider here is whether the monitoring device will be meant only for the implementors of the goals, or whether your team will also want to use this information as a public relations vehicle for others. It might be that more frequent tracking information would be internal only, with highlights or selected mid-goal milestones reserved for an external audience.

Your decisions about audience for the tracking information will determine some of the other possible features of the monitoring vehicle, particularly the **reporting format**. Format refers to how the data will actually appear in a report; that is, as a raw number in a prose sentence, a percentage in a pie-chart graphic, or a number on a spreadsheet. The reporting format will be primarily determined by the audience to which you want the information to go. The raw data exists in an abstract form until it is presented in some kind of analysis; and there are many possible reporting forms that the same raw datum could take depending on the needs and statistical sophistication of the intended audience.

Another aspect to consider is the **reporting frequency** of the measurement. It can be calculated every week, but will it be reported every week? (But, if the data will not be reported every week, does it need to be calculated every week?) It does not necessarily follow that the tracking frequency and the reporting frequency need to be the same; but if there is a discrepancy, there should be a good reason for it. These aspects are obviously interrelated, as the audience you choose to report to will determine, to some extent, the reporting format and frequency.

The **report vehicle** is the generalized name for the communications document in which the statistical information will be delivered to its audience. You will need to formulate answers to the other considerations before you can determine which communications vehicle will be the most appropriate for your needs. If your audience is your core implementation team members, they might choose an email listserv as the best way of getting regular monitoring updates. Other interested colleagues might want to read information on the lunchroom bulletin board. Board members will not have access to the bulletin board, but they will read with interest the company's quarterly meeting notes.

Other considerations might include the resources needed to produce the monitoring information. Will utilizing a "Planning Updates" column in the monthly employee newsletter involve the time of one of the Project Implementation and Monitoring Team members? Or can the current newsletter editor do it? Will the person responsible for gathering be able to distribute the pie-chart graphic online or will there be crossover software problems? The cost of distributing the information might be an issue as well.

All of the aspects of monitoring that have been discussed will need to be determined before a final communications vehicle can be chosen. In fact, there will no doubt be a variety of communications vehicles needed to handle the different audiences and reporting requirements of your project. Your team will have special knowledge of what exists in your unique environment, but some possible reporting vehicles to consider might include the following:

- Email
- Ebulletin posting (or internal listserv)
- Chatroom
- Old-fashioned bulletin board
- Memo
- Company newsletter
- Board Members newsletter
- Quarterly executive meeting notes
- Enclosure in paychecks
- Verbal announcement at department or project team meetings
- Annual report
- Quarterly video updates

All of the aspects of reporting your monitoring information are obviously related. The reporting vehicle—company newsletter, for example—can be used to determine the tracking frequency. The point here is to consider these aspects and choose a monitoring method that makes sense for the goal.

The following exercise takes you and your planning team through the process of making a final evaluation of the measurability of your goal, strategic actions, or subactions; and determining the appropriate monitoring vehicle based on a discussion of the terms outlined above.

EXERCISE 13: HOW DO YOU MEASURE SUCCESS?

Materials: most recent version of the goals and strategic actions by cluster, flip charts, tapes and markers, visual reminder of the aspects of monitoring (Figure 9.1).

Duration: 1–1½ hours.

Objective:

1. Discuss the aspects of effective monitoring.
2. Recheck that the goals are measurable.

3. Devise an appropriate or monitoring vehicle for each goal or goal component based on the application of the effective monitoring aspects.

Begin the meeting with a brief discussion of the aspects of monitoring. If there are other ideas or monitoring suggestions that the team wants to add, capture those on flip charts for posting in the meeting room (20 minutes).

Then assign team member pairs to each one of the goals and strategic action clusters in order to accomplish objectives 2 and 3 (35 minutes).

Have each pair report back in large group the results of their work, emphasizing concerns, issues, or questions that were raised (30 minutes).

Give notes for this session to the wordsmithing group for inclusion in the final planning document.

SAMPLE PLANNING ITEMS: ADDING MONITORING TOOLS

In carrying through on the list of sample planning items, after completion of the above exercise, the goals and strategic actions might be joined with monitoring tools in the following ways (note that R.P. = responsible person):

A. Increase our total market share by 8% in 2 years, to be tracked quarterly, by increasing sales in our existing three-state region by roughly 1% per quarter beginning 2Q 20XX. (R.P.) (Base measurement: total market share as of year-end 20XX.)

Monitoring: as specified above will be computed and compiled by R.P. and passed on to R.P. for the following distribution:

- Intranet: "Planning Goals," posted the last month of every quarter. (R.P.)
- Meeting notes: included in a special tracking section of division managers meeting notes. (R.P.)
- Email: posted via email to division heads as soon as available. (R.P.)
- Email: to all project team members. (R.P.)

Guiding Assumptions

Core competencies focus on people skills and strong branding, not technology or low-price leadership.

Therefore, growth will be achieved by deepening relationships with current customers and by partnering with them to gather intelligence in order to increase our customer base.

Major thrust will be a renewed customer service orientation in preparation for potential new product design and launch (beginning 4Q 20XX).

No dramatic monetary offers will be utilized at this time, except the perceived value of courtesy and closer personal contact.

Strategic Actions/Tasks

1. Customer intelligence survey team tasks include the following:
   a. Identify survey research and design team by 4Q 20XX. (R.P.)
   b. Review existing marketing data on the following items by 1Q 20XX: (R.P.)
      – Customer preferences for current products: sales profile
      – Customer suggestions for new products
      – Reasons for customers' choosing Competitor X's product
   c. If existing data is insufficient, design survey vehicle and implementation plan to capture needed marketing data by 2Q 20XX. (R.P.)
   d. Assemble production team to research printing and distribution mechanics to have survey in customers' hands by mid-1Q 20XX. (R.P.)
   e. Develop analysis tools to evaluate current customers and understanding them using finer demographic customer profiles, based on survey results, 2Q 20XX. (R.Ps.)
   f. Follow up with "We Need You" calls for portion of customers who do not return surveys by end 3Q 20XX. (R.P.)
   g. Research the feasibility of implementing customer surveys for 5% of total current customer base on a quarterly basis. With a go/no-go answer by 3Q 20XX. (R.P.)
   h. Hire mystery shoppers to shop in areas of projected expansion and reporting back to marketing implementation team by 2Q 20XX. (R.P.)
2. Marketing campaign redesign and execution team tasks include the following:
   a. Develop 3Q marketing campaign with a diverse array of approaches—including initial contact vehicle and series of follow-up tools—to match unique, customer-specific demographic niches. (R.Ps.)
   b. Research utilization of Web site for partnering program using permission marketing techniques, beginning 2Q 20XX. (R.P.)
   c. Consider new series of testimonial ads in target market area, mid-2Q 20XX. (R.P.)
   d. Pursue partnership with retail outlets to institute "Open House" tables and demonstrations at local outlets with the purpose of establishing personal contact with our customer base, mid-3Q of end 4Q 20XX. (R.P.)
   e. Consider the formation of "You Tell Us How" teams of consumers to advise marketing group, in exchange for special customer offers, "Thank You" gatherings, and appreciation awards, 2Q 20XX. (R.P.)

3. Training design and delivery team tasks include the following:
   a. Co-design, with sales, generalized training that reflects new tactics and attitudes about customers, 1Q 20XX. (R.P.)
   b. Work closely with marketing campaign team to design suitable training for new programs, goals TBD.

Potential Cross-Team Issues

Continue to nurture closer working relationship between marketing and sales: form joint implementation team?

Need support from training department on customer care training design and delivery?

Need to work with personnel on hiring criteria to reflect new attitudes?

B. Internal challenge: increase the success of our sales campaigns, measured against rate of sales conversions of responders (using 1Q 1998 campaign as a baseline) with specific goal TBD based on campaign targets set by marketing team. (R.Ps.)

Monitoring: responder rate data will be computed and compiled by R.P. daily during both the pre-holiday and spring campaigns and distributed by R.P. as follows:

- Marketing bulletin board by the copy machine by 9:00 AM for previous day. (R.P.)
- Email posting to all marketing/sales team members by 10:00 AM for the previous day. (R.P.)
- Aggregated and reported at weekly department meetings for marketing, sales, customer service and fulfillment. (R.Ps.)
- Aggregated and reported in bi-weekly status reports to division heads. (R.P.)
- Final results included in quarterly status report to Board. (R.P.)

Guiding Assumptions

We believe the weaknesses in the current sales campaigns are the result of two general factors: ineffective campaign execution and the need for a revised campaign approach.

Specifically, we want to improve the following: (1) the breakdown of internal communications between marketing and sales; (2) inadequate follow-up after the initial customer contact has been made; and (3) a 'one-size-fits-all' approach to the customer (note crossover with overall marketing goal A above).

Strategic Actions/Tasks

1. Marketing/sales joint task force: establish a marketing and sales support team by 1Q 20XX—including sales, customer service, fulfillment—with the following objectives: (R.P.)
   a. Review current hand-off practices for sales campaigns, 1Q 20XX. (R.P.)
   b. Consider and/or design special program for new customers: welcome call and survey, and periodic check-in mailers, 3Q 20XX. (R.P.)
   c. Design and delivery of training for internal personnel who handle new accounts, 2Q 20XX. (R.P.)
2. Email tech team tasks: support/initiate project for the enhancement of headquarters-to-branch email capacity, Jan 15, 20XX. (R.Ps.)
3. Web site task force task: assess computer sophistication of current customers and, if warranted, meet with IT team to discuss the marketing and distribution potential for extranet Web site, 2Q 20XX. (R.Ps.)

Potential Cross-Team Issues

IT and Marketing/Sales initiatives will require cross-team support.

IT may also need to be involved in current customer computer-use analysis.

Will there be shared budget issues on potential extranet initiatives?

C. Department challenge: reduce paper use in the department by 15% and copier usage by 10% by 3Q 20XX (R.P.) and expedite new contract account set-up from 8 weeks to 3, by end-of-year 20XX (R.Ps.)

Monitoring: track paper and copier usage by month and report as follows: (R.P.)

- Marketing bulletin board by the copy machine by the first Tuesday following month-end. (R.P.)
- Aggregated and reported in bi-weekly status reports to division heads. (R.P.)

Monitoring: progress on new contracting procedures will be documented in weekly status reports by R.P. to department heads.

Guiding Assumptions

The assumption is that increased paper and copier usage can be combatted by environmental education and a focus on environmentally-friendly practices in the department.

It is assumed that legal will work with us to consider and improve the efficacy of contracting procedures.

Strategic Actions/Tasks

1. Contracting procedures task force: in collaboration with legal, review contracting procedures in order to reduce the amount of time for new contract account set-up from eight weeks to three, by end-of-year 20XX. (R.P.)
2. Reduce paper use task force:
   a. Open broader discussion with department members about preferred communications modes: paper, electronic/digital media, face-to-face, etc. (At next department meeting.) (R.P.)
   b. Suggest at next quarterly division head meeting that a task force be appointed and a series of environmentally-friendly policies be developed, 1Q 20XX. (R.P.)
   c. Research national or local environmental organizations to see what material already exists, 2Q 20XX. (R.P.)
   d. Research broader applications of an environmental policy: copier cartridges, laser-jet ink replacements, recycling of all office waste materials, etc., 2Q 20XX. (R.P.)
   e. Disseminate information and begin campaign by 3Q 20XX. (Wrap decreased copier and paper usage goals inside the corporate campaign.) (R.P.)

Potential Cross-Team Issues

To be successful, the environmentally-friendly policies must be adopted corporate-wide.

SUMMARY

By this point your team should have most of the elements of a Strategic Plan in place: challenges (in clusters), goals (measurable and assigned), strategic actions (with guiding assumptions and activities), and monitoring details (with measurement format, frequency, and communications vehicles).

Congratulations!

Now you just have to make sure the plan is followed. But do not breathe a sigh of relief yet—the execution of your plan will provide its own set of challenges and frustrations. So, what is that expression?—no rest for the weary?

Chapter 10 focuses on some implementation tips that should help in making a smooth transition from the process of planning to the procedures of doing.

NOTES

1. Hubbard, Justin, "Knowing What We Know," *InformationWeek*, November 17, 1997, 100.

2. For more information on Interface Inc., the parent company of Bentley Mills (one of the 26 factories in the company), and the incredible innovations that CEO Ray Anderson is instituting there, see *Fast Company*, April: May 1998, an article entitled "The Agenda," by Charles Fishman, p. 136–142; or see the Profile in *InnerEdge* (Vol. 1, No. 2, June/July 1998) on Ray Anderson written by David R. Merkowitz, p. 25–26.

3. Unfortunately, this book by authors Peter Lorange and Richard F. Vancill, *Strategic Planning Systems*, Prentice-Hall, Englewood Cliffs, NJ, 1977, is currently out of print, although it is still available in some libraries. The quotation cited appears on page xiii in the introduction of this book.

4. Information taken from the *American Heritage Dictionary*, CD version, produced by WordStar International, licensed from Houghton Mifflin Company, copyright 1993.

III

POST PLANNING

10

IMPLEMENTATION TIPS

Well begun is half done.

Horace[1]

OVERVIEW

It sounds too absurd to be true, but it is almost a sure bet that most of you can tell a story about a planning process that took months, produced hours of work and weeks of meetings for everyone, and resulted in a prominently displayed document that collected dust on a shelf. Maybe the resulting plan was even a good one: well-conceived, innovative, and timely. But nothing was ever implemented.

Even if a plan is not completely ignored, too often after it has taken shape, the team spirit and all the team interactions that worked so well during the planning phase come to an abrupt end—almost as if someone had turned off a switch. The minute the plan has been approved by the right people, the entire structure that has supported the process of its creation is unceremoniously disassembled. Everyone begins "doing" the plan. No more meetings. No more discussion. Everyone is on their own. And, not surprisingly, many of the planning goals are never met.

Perhaps it would be good to think of the plan itself as the halfway mark, rather than a product signifying an endpoint. The plan is only a map; getting to your destination is the goal.

As discussed, there are many reasons why a good plan does not produce the expected results. But for *your* plan and *your* planning process, do not let one of them be poor implementation follow-through. This chapter offers some suggestions for making sure that your well-crafted plan does what it is supposed to do—produce the desired results.

THE PLANNING DOCUMENT

Despite its bad reputation as something to prop the door open with, there are many good reasons for completing a formal planning document. For one thing, it is nice to finally have a product after so many hours relegated to process. The plan is a summing-up of your efforts, and it captures, for at least one point in time, your team's thinking and analysis. This product is something that you and your team members can refer to, show to others, even keep as a portfolio item when it is time for the next job search.

Just as the audience is one of the first aspects to consider in selecting a monitoring and communications vehicle, here, too, one must look carefully at who might want to see and use your planning document. The considerations about how the document will be used and who will read it will assist you in understanding how best to format and produce your plan.

Depending on the needs of your organization, you may want a document that can serve as its own communications device, both within the organization and to be used selectively with other audiences: Board of Directors, funding sources, potential employees, or the media.

Possible audiences for a planning document might include:

Internal:
- Board of Directors/shareholders
- Executive Management team
- Planning Team members
- Other Team implementors
- Potential employees

External:
- Customers
- General public
- Financiers/bankers/venture capitalists
- Analysts, industry SMEs (subject matter experts)
- Media

If planning to use your document for more than one major audience group, you will probably want to have different versions of it. For example, you might not want to share competitive information with the media or members of the general public; whereas, that would be information of most interest to your Board Members and other financial stakeholders. This multi-version approach might argue for a loose-leaf binder that could be put together on a just-in-time basis from a full range of document sections and pages.

If you choose to make different versions of the document, immediately determine a method of version control that can be imprinted on every page; date and time is an easy one. In my experience, the text of a planning document will quickly become a moving target as conditions change and corresponding changes are made to the document. For that reason, there should be one person assigned to manage your document—to provide the 'portal of authority' for the document—to keep track of what changes are being made and who is receiving which versions.

Regarding crafting the words on the page, it is my opinion that group writing is generally a disaster. So pick a wordsmith or a hired-hand—or maybe there is a planning team member who can take the job—and pass off all current notes to the designated writer so that he or she can put your planning information into a standard format. If the writer was not present for the planning process, be sure that you assign someone who was present to be the SME to answer questions or clarify ambiguous information. Expect the finalization of the text to take many rounds of editing and approval.

Keep it short, or provide an executive summary that highlights your major assumptions and goals. If there is background information or supporting data that you think ought to be included, put it into an appendix where it will be available to those who want detail but not be in the way for those who would rather cut to the chase.

Be sure to give credit where credit is due—prominently list all your planning team members and all senior or executive team members who supported your efforts. Do not forget to thank caterers and support staff, even if it is only a small mention. The more you can draw people's interest and attention to the document, the better it will serve you.

Not a lot of money needs to be spent to make your document a showpiece; but a nice inexpensive binding will make the plan durable and usable. If your organization has a rudimentary in-house print shop, your plan's production can be cost-effective. Perhaps you even have an artist on staff who can design a cover. If not, it is still worth getting a selected number of copies professionally reproduced and bound: either a heat-tape binding, or a drill-and-press bindery process is relatively low in cost, good-looking, and durable.

With any luck, you will have a document that will find many uses as the implementation process proceeds.

THE COMMUNICATIONS PLAN

One of the most important aspects of implementing any planning project is communicating about the project itself—how the plan was conceived, and what will take place during the project's execution. A communications plan is simply the result of a mini-planning process that answers the question—who needs to know what and when? You might think of a communications plan for your strategic plan as internal PR for your planning project.

The communications plan is important but, depending on your level of expertise and interest, you should keep it simple. Probably a few members of the planning team can get together and rough out a communications plan, but it will take some thought. Do not skip this step or a lot of the hard work put into the planning process will be lost. Do not lose control of what people know about your plan. It might be a good idea to consult with or to bring onto the communications planning team a member of your corporate affairs department—someone who knows the entire organization and may already have in place many of the communications distribution pathways that your group will need to disseminate planning information.

A kick-off announcement about the completion of the strategic planning process is often a good idea—especially if there is an event already on the calendar that can be commandeered (or shared) for that purpose. As mentioned before, event-driven deadlines are always the most effective. A lot of energy can be created in preparation for something that must happen on a specific day, or else. A kick-off allows you to both spread the word about what your group has accomplished in a fun way as well as reward your team members for all their hard work.

The event could be the bi-weekly division-heads' meeting, the quarterly luncheon with Board Members, or the annual employee picnic. Whatever it is, find a way to create a celebratory atmosphere. Buttons, balloons, and baseball caps might seem hokey, but everyone enjoys them and you will find a lot of these tshatshkis gracing employees' desks and cubicles years after the event. Have a photographer on hand to catch some casual shots of the event. (Maybe take a few pictures of your planning team: a group shot scanned into a graphics digital file could be incorporated into the planning document cover.) If it seems appropriate, make a formal presentation of the strategic plan to the CEO, COO, Director of the Board, or even a department head. Ceremony is about creating events that you imbue with meaning; so be creative.

When I was a trainer at CitiSource (the name of our training department at Citicorp, Western Division), we created a "helping hand award" that was given out every quarter to one member of our department who had been particularly helpful to others. It was simply a pink glove, stuffed with beans that sat upright on a stiff collar, with its fingers wiggling. But the way it was presented—all managers were present to shake the helping hand of the recipient—was funny and memorable. Many of my corporate mementos have been laid to rest in the circular file, but I still have my pink glove.

In addition to identifying a kick-off or initial announcement, your communications plan should address how you will update concerned people about the progress made toward the goals in your plan. The employees most directly involved will probably be reached by the various monitoring vehicles that you have designed into your process, but you may want to take into account some types of announcements that would be more appropriate for a larger, general audience. Is there a national corporate newsletter looking for a story? Can you get a mention in one of the annual corporate videos? Is your planning process unique enough to be written up as part of another professional organization's journal? If you will be communicating major milestones with a larger audience, you may need to create a process to aggregate some of the monitoring information to target a few key goals.

The main point about a communications plan is not to leave anybody out and to make sure that, if people get information in stages, you have the right sequence of announcements. The CEO should not be the last to know.

THE IMPLEMENTATION TEAM

As mentioned in Exercise 12: On Target Action/Final Check, one of the last tasks for the entire planning team is to appoint key planning team members to be part of the Implementation Team. If you have ended up with several groups of implementing teams working on different challenge clusters, it would make sense to include one representative from each of these teams. Or if your implementation groups are divided by division or department, then appoint one member from each department.

The Implementation Team's task will be to guide the implementation of the strategic plan by providing a central body to oversee matters of concern to the planning execution as a whole. Its various tasks or roles could include the following:

- Resolving conflicts or mediating cross-team issues
- Lobbying upper management for needed resources or providing liaisons to upper management for removal of any barriers to project success
- Coordinating budgets and interdependent timelines
- Gathering and evaluating all tracking/monitoring information
- Troubleshooting if plan begins to get off-track, based on predetermined flashpoints for strategic action direction changes

■ Providing a consistent, centralized committee for all planning-related activities or questions

Do not let this job get pawned off on weak members of the team. The make-up of members will determine the strength of this team and its efficacy; and the efficacy of the Implementation Team may well determine the ultimate success of your planning project. If you have some influence over member selection, make sure that the key influencers, or the movers and shakers of your planning process, are on this Implementation Team. Not only will the team members give the project the prestige it will need to be successful, but it will provide some heavyweights if, or more likely when, the implementation process hits the inevitable snags.

One aspect of the strategic planning process that has not been explicitly covered by this methodology is the establishment of a budget for the planned strategic actions. It is assumed that planning this budget can be part of the regularly scheduled operational budget-building exercise that occurs in most corporations. If that is not the case, budget coordination will be most easily handled by this Implementation Team.

The team may also want to consider, before the need arises, how to facilitate agreement on a process for trouble resolution if tasks remain undone or if hand-off of linked strategic actions do not happen smoothly. In terms of administrative duties—meeting chair, minutes note taker, minutes distributor, etc.—it is a good idea to rotate these responsibilities so that the burden of these jobs is spread evenly among all team members.

The last order of business should be to determine the frequency of Implementation Team meetings and to publish a meeting calendar for the entire course of the planning timeframe you have decided on: six months, eight months, one year, etc. As anyone who works in a corporate environment knows, finding a meeting time for all the people who need to be at a meeting has taken on the complexity of rocket science. The best advice I can give is to establish a standard meeting time—every other Wednesday at 9:00 A.M., for example,—and stick with it.

THE INFORMATION AND COMMUNICATIONS MÖBIUS

The Information and Communications Möbius is a concept introduced as a metaphor for how the Implementation Team can be most efficient and empowered to do its job. The basic planning cycle was introduced in Chapter 2 as if it were a loop or spinning triangle. Here, that planning cycle loop is augmented to incorporate other spatial elements.

The Möbius metaphor (Figure 10.1) is a relative of the concept of Total Quality Management (TQM) introduced initially by W. Edwards Deming

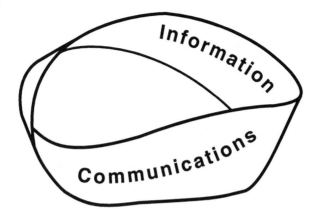

Figure 10.1   The Information and Communications Möbius

to control errors in manufacturing. The Information and Communications Möbius is a way of talking about continuous output and continuous feedback; it stresses the need to have quality information about progress on goals available in timely, continuously running cycles, to evaluate or analyze the information accurately, and to deliver it back to team members efficiently and quickly.

Take a moment to make a Möbius strip and you will understand the power of this metaphor. It is quite simple. Grab a piece of 8½" by 11" paper and slice off a strip about ½" wide down the long side of the paper. Now you have a strip ½" wide and 11" long. It has two sides, clearly. On one side of this strip in the middle, write "Information." On the side of the strip, just opposite "Information," write "Communications." Now take the two ends of this strip and, instead of connecting them together as if you were going to make a ring, make one twist in the strip and tape the two ends together.

Now you have a Möbius strip—a continuous ring of paper with *only one side*—and the concepts of information and communication have become linked on one continuous surface. Don't think so? Just take two fingers, pinch the strip and slide along the surface of the strip. You will not have to lift your fingers to make it entirely around the ring, passing both "Information" and "Communications" conjoined on one surface.

So what does this mean? It is an image for the importance of having quality information continuously available, interpreting that information accurately, and distributing or communicating that analysis back to your team in an uninterrupted stream. Is it possible to do that? Your body does it every moment of every day.

Each millisecond of every waking minute, the hair-like nerve endings in the cochlea of your inner ear send impulses to your brain. These are then coordinated with other signals from the eyes and sent back out as feedback to the 60% of the body's muscles devoted to balance.[2] It's the monitoring team called, in this case, the brain, that takes in information, analyzes it, and sends it back to the implementation team at large (the muscles) that makes things work.

I have seen two or three companies that have been able to put something close to this concept into practice. Manifestation can take many forms, depending on the actual structure or make-up of a company. One large corporation uses a robust intranet linked in such an intricate way to employees' everyday business that it is automatically and categorically updated with information as soon as the information has been created. This system handles the "information" half of the Möbius but omits the communications half. To complete the loop, the system would need to be able to post back to users, instantly and automatically, information relevant to their various projects or inquiries: something similar to the so-called "pull" concept on the Internet, wherein users can establish categories of information of interest to them for retrieval and presentation on their desktops.

Is there a modified version of this mechanism that your Implementation Team can put into effect? The process of creating and communicating monitoring devices is a start. Talking at the water cooler is a start. Any formal or informal system that can be created for establishing lines of communications about the project will become part of the Möbius. Your Implementation Team will need to be on the Möbius or will need to provide a support environment for its functioning.

At any rate, this image is offered here simply as an aid for conceptualizing how a well-functioning systems works. If a system is working well, it will probably resemble the body or a colony of ants. The more one looks to biological models for an understanding of best (best in the sense of most efficient, self-organizing) practices, the more smoothly business organizations will run.

THE LAST WORD

"Implementation" is an interesting word with many complex derivations. The American Heritage Dictionary says, "The verb implement, meaning 'to put into practice, carry out,' has been in use since the 19th century. Critics have sometimes objected to the verb as jargon, but its obvious usefulness appears to have outweighed their reservations."

One of its Latin roots, *pelo*, means full, fill, or plenitude and, even further back in time, a Sanskrit derivation exists with the secondary meaning of "cake"—or "that which satisfies."[3]

In fact, conceiving a strategic plan without the satisfaction of implementing it does seem a particularly empty activity. I hope these implementation tips will fill your team members with the enthusiasm they need to complete the second half, and now, the most important part of the job.

NOTES

1. Horace, *Epistles* I. ii.
2. The description of walking is taken from an article entitled, "A Delicate Balance," by Barry F. Seidman, published in the May/June 1998 issue of *The Sciences*, p. 12.
3. Information taken from the *American Heritage Dictionary*, CD version, produced by WordStar International, licensed from Houghton Mifflin Company, copyright 1993.

11

FUTURE PERFECT

We are in free fall into the Future.

Joseph Campbell[1]

In the text accompanying this quotation, Joseph Campbell talks about the pace of change in what he calls life-conditions in our contemporary world. There is a huge lag time now between the conditions reflected in our society—legal practices, political legislation, family and church values—and the actual realities of life. He describes it as a grand adventure—this free fall—almost as if it were a ride at a theme park:

> We can no longer hold on with confidence to the system which once worked—the country has got to open now to other things.[2]

The business world is exciting because this lag time is shortened; it has to be. If a business is not responding effectively to conditions in the real world, it will die, sometimes quite quickly. It did not take Barnes and Noble long to realize they had to figure out how to compete with the incredible success of that upstart amazon.com. The era of no brick-and-mortar businesses is here. In the business world, the future is now.

It is this future that will unfold to reveal the ultimate results of your planning project: were your business assumptions correct, and were the strategic actions your team devised the right ones? In a metaphoric sense, the future is perfect in that what is about to happen in it is already perfectly formed; it is only a fraction of a moment away from becoming the present. Yet, it is the future that holds the seeds of developments that no one has anticipated. Discovering the secrets the future holds is an exciting adventure, and all you need to do is wait!

Before you know it, your project objectives will be accomplished and will be replaced with others. But before the events of this recent planning project get too far into the past, bring your team together for an evaluation meeting. In this era of the fleetingness of time, it might be even more important than it used to be to pause occasionally to reflect on what is happening or what has just happened in order to come to a greater understanding of where we are. There is no lack of information in the information age; but what is often missing is analysis.

At this point in your planning process, no one will want one more meeting than is necessary. But a final evaluative get-together will provide information that you cannot obtain in any other way. You and your team have worked hard together and you owe it to yourselves to discuss exactly what you have done and how you have done it.

The exercise that ends this chapter is meant to be completed as an evaluation for the planning process itself. The plan implementation, of course, will still be on-going and many of the same team members will be involved in those efforts. But the planning portion of the project has come to an end. One of the unspoken reasons for this meeting is to give your team a sense of closure on that process. Acknowledging what you have done well and discussing where improvement could have been made will allow your team to let this process go and to begin the next phase: the implementation of your project.

EVALUATING YOUR TEAM'S PLANNING PROCESS

You probably have access to a variety of evaluation forms for a variety of purposes. Maybe there is a standard evaluation form used in your organization. Take a look at what is available to see if it is suitable for evaluating your team's planning process. If not, you may want to create an evaluation of your own using some of the sample questions listed below.

This evaluation process is not meant to yield hard statistics. After all the discussions about requiring goals to be quantifiable, in fact, one must admit that there are times when soft information is preferable. This is one of them. I prefer an evaluation form that encourages thinking and discussion rather than one with boxes to check. Depending on how you want to handle the evaluation, you may want to hand or send out this form to team members before they come to the meeting. At the very least, team members can scan the form and be thinking about the issues and questions raised. In general, however, you will have better luck actually getting it filled out by giving people time to do so as part of your meeting.

The exercise encourages a group discussion process after the questions are answered in writing by each individual. Since this project has been a group planning effort, this group discussion seems appropriate. But, depending on your group, you may not get the level of honesty and frankness that you might if the form were completed and handed in to you privately. Use your discretion in this matter, depending on what information you think might be forthcoming.

Here are some sample questions to consider in putting together your own evaluation form:

Planning Process

- Did the process differ from your expectations for the process? If yes, how?
- If yes, were the differences OK, difficult to accept, unneeded? What bothered you about them or what did you like?
- Was the structure of the process OK? Was the length of the meetings appropriate? Did the exercises accomplish their purpose? Was the process explained clearly?
- Did the process accomplish what it set out to accomplish?
- What worked particularly well?
- What should we change to improve our process next time?

Team

- Did we have the right representatives on the core planning team?
- Did the team abide by the agreed-to rules? Or did the rules and agreements need to be revised along the way? Comment.
- Did you feel that your contributions were appreciated? Incorporated?
- Were disagreements resolved in a satisfactory manner?
- If disputes arose, did participants feel comfortable enough to continue with the process? If not, why not and what changes would you recommend?
- Overall, do you think this team planning process was successful? Why or why not?

Self

- In the course of participating in this process, what were your major learnings?
- What do you feel was your strongest personal contribution to the planning process?
- What was the greatest challenge that you faced in the course of this planning process?

■ What advice would you give to a team leader or participant for the next planning process?

EXERCISE 14: HOW DID WE DO?

Materials: flip charts and markers; planning evaluation form; copies of your original Planning Process Agreement.

Duration: 60 minutes.

Objective: to evaluate the team's planning process; to come to closure.

Hand out evaluation forms with questions selected from the list above. Give individual team members time to complete their answers on paper (15–25 minutes). Then open the discussion to the large group. Encourage those who may have a more personal or private comment to speak to you directly, or, if appropriate, to hand in their forms anonymously.

Revisit the agreements that were made in Exercise 1: Creating Agreement. Review your contract with one another. Did the team abide by the agreed-to rules? What needed changing?

The simplest way to organize information on your flip charts is to chart in two categories: "What we liked/Did right" and "What we would like to change." These items can be kept for review as part of the new process—for example, as part of completing another Exercise 1: Creating Agreements for the next team—or they could be handed off to the next planning team leader.

Perhaps you might end the meeting by sharing any personal learnings you have gained as leader of the process. Others may then feel comfortable enough to share their own learnings.

As a closing, thank your team. Or read a poem. Or whatever seems meaningful and appropriate.

FUTURE PERFECT

And now, let it go.

The future perfect is a verb tense in English that expresses action that will be completed at a certain time in the future. What is unusual about this verb tense is that it is formed with the past participle of the verb, something generally only used for past tenses, and a future helping verb.

So, you might say:

We *will have done* a superb job on our planning project by the time you read this sentence next year.

Make it so.[3]

NOTES

1. Quote taken from *An Open Life: Joseph Campbell in Conversation with Michael Toms*, Selected and edited by John M. Maher and Dennie Briggs, Larson Publications, 1988, 87.
2. *ibid*, Campbell, 87.
3. Favorite expression of Jean-Luc Picard, captain of the Starship Enterprise, in the Next Generation Star Trek series.

POSTSCRIPT

I hope I have created a text flexible enough for use by those of you in traditional corporate settings, fast-paced start-ups, nonprofits, government agencies, or even small businesses; but once the book leaves my hands at this keyboard, how and where it is used will be up to you. I am curious, of course, whether you have used any of the exercises or only the exercises, scanned the resource listings, or copied out only a quotation to send to a friend.

Being a writer is an ironic occupation for a consultant. I have sat in front of my computer and created systems and sentences for several months now without the benefit of any input at all from you, the reader. And although this book is about teamwork, and my preference is to work in collaboration with clients and co-workers, the writing process is primarily solitary.

But perhaps the dialogue can begin now.

If you have found the book helpful, or wish something had been done differently in producing it or have been confused by some part of the planning method, I encourage you to let me know by emailing me at axioun@aol.com. I am anxious to know who you are, how you used this text, and what I can do to make the ideas in this book more relevant and useful to you in future editions.

Let me leave you with two of my poems: one published by Axioun Books in 1976 in a collection entitled *Gatherings*; the other unpublished until now.

each word's
all unspoken others'
contained absence.

Mind
The field.

That should speak to Jacob Bronowski's concerns and cover whatever I have left out here.

APPENDICES

APPENDIX A

List of Exercises

Exercise # and Name	For Whom?	Page #	Chapter
Exercise 1: Creating Agreement	Team	12	Chapter 1: My Team, My Self
Exercise 2: Influences from the Past	Leader	30	Chapter 2: Beginning Concepts
Exercise 3: Influences from the Future	Leader	37	Chapter 3: Becoming a Strategist
Exercise 4: Nurturing Your CHI	Leader	44	Chapter 3: Becoming a Strategist
Exercise 5: What Needs Fixing?	Team	70	Chapter 5: Seeing the Challenges
Exercise 6: Alignment with Management	Optional	72	Chapter 5: Seeing the Challenges
Exercise 7: Clustering the Challenges	Team	82	Chapter 6: Sorting the Challenges
Exercise 8: Prioritize and Finalize Challenges	Team	89	Chapter 6: Sorting the Challenges
Exercise 9: Transforming Challenges Into Goals	Team	100	Chapter 7: Setting the Goal
Exercise 10: Influences Reprise	Team	115	Chapter 8: Devising Strategic Action
Exercise 11: Intentional Insight	Team	118	Chapter 8: Devising Strategic Action
Exercise 12: On-Target Action	Team	119	Chapter 8: Devising Strategic Action
Exercise 13: How Do You Measure Success?	Team	138	Chapter 9: Monitoring and Measurements
Exercise 14: How Did We Do?	Team	160	Chapter 11: Future Perfect

APPENDIX B: SEQUENTIAL OVERVIEW OF THE CGSM© PLANNING PROCESS

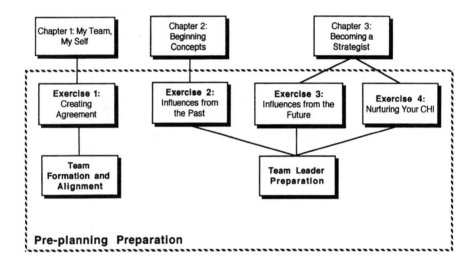

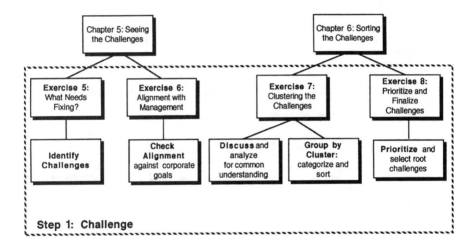

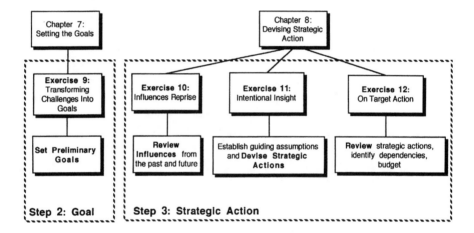

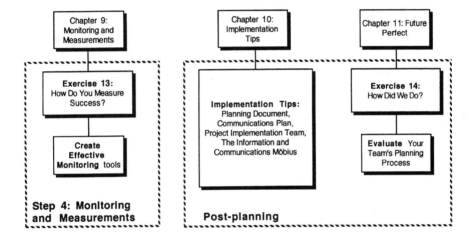

APPENDIX C: SAMPLE PLANNING ITEMS

Challenges ⇒ Goals ⇒ Strategic Actions ⇒ Monitoring

1. BRAINSTORMING CHALLENGES

- ■ √√√√ Bad communications with sales department/sales reps: they always get the offer wrong, not prepared for product changes, etc.
- ■ √ Lack of follow-through with customer service reps, i.e., "bad press."
- ■ Logo and collateral outdated—need new look.
- ■ √√ Need a "cooler" Web site.
- ■ Procedures for hiring outside graphics vendors too cumbersome.
- ■ Competitor X is poaching our best customers.
- ■ Department is down two staff members.
- ■ √ Support staff not responsive enough to deadlines.
- ■ Why not create a new product for the nostalgic Baby Boomers?
- ■ √√ How can we get more timely competitive info?
- ■ √√√ Increase sales revenue to meet team goals.
- ■ √ Re-evaluate market share numbers.
- ■ Get better coffee and real cream!
- ■ Fax too slow.
- ■ √√ Cut down paperwork!!!!!!
- ■ Benefits for significant others?

√ checks indicate the number of times a particular challenge was mentioned by a team member.

2. CHALLENGE CLUSTERS

External Business Challenge/Marketing

- ■ Need new direct marketing campaign to increase visibility in target region.
- ■ Do not know potential new target audiences within our region.
- ■ What do our current customers like or dislike about our product or service? What about new product offerings?

- New collateral to support direct marketing campaign—budget?
- Why are customers leaving for Competitor X?
- Increase our market share in our three-state region by 8% in 2 years.
- ⇓ Do we need a better Web site? including customer FAQs or chatroom?
- ⇓ We need better competitive information at more timely intervals.

Internal Business-wide Challenge/Communications Issues

- Sales department/sales reps too removed from marketing information, need to hear it from "the horse's mouth?"
- Customer service reps overwhelmed by a rash of 'new customer' problems during our campaigns.
- ⇑ Do we need a better intranet Web site? including customer database access, current sales info, etc.?
- Inadequate email system with branches? realtime, not batch, processing.
- ⇓ Do we need so much paper communication internally?
- Benefits for live-in partners possible?

Internal-Department Challenge/Administration

- ⇑ Contracting procedure too cumbersome? Check with legal.
- ⇑ Do we need so much paper communication internally?
- Need faster fax.
- ~~Better coffee~~ solved!

Note that the arrows indicate that a challenge is repeated in the category either just above or just below it in our hierarchy, meaning that it will probably need to be solved or addressed on multiple levels.

Or if your world makes better sense divided into team groupings or functionalities, your sorted challenge list might look something more like this:

IT/Techie Team

- ⇓ Do we need a better Web site? internal: including customer database access, current sales info (see administration); and external: customer FAQs, chatroom, (see Product X team).
- ⇓ Inadequate email system with branches: need realtime, not batch, processing, (see administration).

Administration/Support Staff

- Contracting procedure too cumbersome.
- ⇑ Do we need so much paper communication internally? (See IT.)

Product X Team

- Do not know potential new target audiences within our region.
- ⇑ What do our current customers like or dislike about our product or service? what about new product offerings? (See IT.)
- Why are customers leaving for Competitor X?
- Increase our market share in our three-state region by 8% in 2 years for sales of X.

Creative

- New collateral to support direct marketing campaign—budget?
- Need new direct marketing campaign to increase visibility in target region.

All/Special Projects

- Benefits for live-in partners possible?
- ~~Better coffee~~ solved!

Note ⇑ or ⇓ arrows indicate a challenge either supports or depends on another team's challenge.

3. FINAL CHALLENGES

> **A. Overall external business challenge: increase our market share in our three-state region by 8% in 2 years.**

Supporting Marketing Challenges

1. Need competitive/customer info: why are customers leaving for Competitor X? Do not know what our current customers like or dislike about our products and services.
2. Do we need new product offerings (i.e., to take advantage of the aging Baby Boomer market)?
3. Not taking advantage of potential new target audience within region.

4. Current marketing collateral needs new look—budget?

5. Design new direct mail campaign? Other distribution vehicles?

B. Internal business challenge: improve our communications with other, linked departments, i.e., sales, customer service, fulfillment.

Supporting Communications Challenges

1. Sales department/sales reps too removed from marketing information.

2. Customer service reps overwhelmed by a rash of 'new customer' problems during our campaigns.

3. Do we need an intranet Web site for internal customer database and sales info? Or extranet Web site as new distribution channel? (Marketing tie-in above?).

4. Slow or no email possible between headquarters and branches.

C. Department challenge: department communications inefficient (too paper-based) and contracting not expedient (takes too much time).

Supporting Administration Challenges

1. Contracting procedure too cumbersome—check with legal?

2. Too much paper in our internal communications process? Or how do we want to communicate with one another: paper, electronic/digital media, face-to-face?

4. SAMPLE CHALLENGES AND GOALS

A. Overall external challenge/goal: increase our total market share by 8% in 2 years, to be tracked quarterly, by increasing sales in our existing three-state region. (R.P.)

Supporting Challenges/Goals

1. Review existing marketing data on the following items by 1Q 20XX:
   a. Customer preferences for current products: sales profile.

    b. Customer suggestions for new products (i.e., to take advantage of the aging Baby Boomer market?).

    c. Reasons for customers' choosing Competitor X's product.

2. If existing data is insufficient, design survey vehicle and implementation plan to capture needed marketing data by 2Q 20XX. (R.P.)

3. Research the feasibility of implementing customer surveys for 5% of current customer base on a quarterly basis, with a go/no-go answer by 3Q 20XX. (R.P.)

4. Design and implement marketing programs or new collateral material in support of overall market share; goal TBD.

B. Internal challenge/goal: increase the success of our sales campaigns, measuring the rate of sales conversions of responders (using 1Q 1997 campaign as a baseline) with specific goal TBD based on strategic action program development. (R.Ps.)

Supporting Challenges/Goals

1. Establish a marketing and sales support team by 1Q 20XX—including sales, customer service, fulfillment—with the following objectives:

    a. Review current hand-off practices after new sales campaign is launched.

    b. Consideration of special program for new customers: welcome call and survey, and periodic check-in mailers.

    c. Design and delivery of training for our internal personnel who handle new accounts.

2. Meet with IT team to discuss the marketing and distribution potential for extranet Web site and for the enhancement of headquarters-to-branch email capacity, Jan 15, 20XX. (R.Ps.)

C. Department challenge/goal: reduce paper use in the department by 15% and copier usage by 10% by 3Q 20XX (R.P.) and expedite contracting procedures (with goal TBD).

Supporting Challenges/Goals

1. In collaboration with legal, review contracting procedures in order to reduce the amount of time for new contract account set-up by end-of-year 20XX. (R.Ps.)

2. Open broader discussion with department members about preferred communications modes: paper, electronic/digital media, face-to-face, etc. (R.P.)

5. SAMPLE GOALS AND STRATEGIC ACTIONS

> **A. Increase total market share by 8% in 2 years, to be tracked quarterly, by increasing sales in our existing three-state region by roughly 1% per quarter beginning 2Q 20XX. (R.P.)**
>
> **(Base measurement: total market share as of year-end 20XX.)**

Guiding Assumptions

Our core competencies focus on people skills and strong branding, not technology or low-price leadership. Therefore, growth will be achieved by deepening relationships with current customers and by partnering with them to gather intelligence in order to increase our customer base.

Major thrust will be a renewed customer service orientation in preparation for potential new product design and launch (beginning 4Q 20XX).

No dramatic monetary offers will be utilized at this time, except the perceived value of courtesy and closer personal contact.

Strategic Actions/Tasks

1. Customer intelligence survey team tasks include the following:
   a. Identify survey research and design team by 4Q 20XX. (R.P.)
   b. Review existing marketing data on the following items by 1Q 20XX: (R.P.)
      - Customer preferences for current products: sales profile
      - Customer suggestions for new products
      - Reasons for customers' choosing Competitor X's product
   c. If existing data is insufficient, design survey vehicle and implementation plan to capture needed marketing data by 2Q 20XX. (R.P.)
   d. Assemble production team to research printing and distribution mechanics to have survey in customers' hands by mid-1Q 20XX. (R.P.)

e. Develop analysis tools to evaluate current customers and understand them using finer demographic customer profiles, based on survey results, 2Q 20XX. (R.Ps.)

f. Follow up with "We Need You" calls for portion of customers who do not return surveys by end 3Q 20XX. (R.P.)

g. Research the feasibility of implementing customer surveys for 5% of total current customer base on a quarterly basis. With a go/no-go answer by 3Q 20XX. (R.P.)

h. Hire mystery shoppers to shop in areas of projected expansion and report back to marketing implementation team by 2Q 20XX. (R.P.)

2. Marketing campaign redesign and execution team tasks include the following:

a. Develop 3Q marketing campaign with a diverse array of approaches—including initial contact vehicle and series of follow-up tools—to match unique, customer-specific demographic niches. (R.Ps.)

b. Research the utilization of Web site for partnering program using 'permission marketing' techniques, beginning 2Q 20XX. (R.P.)

c. Consider new series of testimonial ads in target market area, mid-2Q 20XX. (R.P.)

d. Pursue partnership with retail outlets to institute "Open House" tables and demonstrations at local outlets with the purpose of establishing personal contact with customer base, mid-3Q of end 4Q 20XX. (R.P.)

e. Consider the formation of "You Tell Us How" teams of consumers to advise marketing group, in exchange for special customer offers, "Thank You" gatherings, and appreciation awards, 2Q 20XX. (R.P.)

3. Training design and delivery team tasks include the following:

a. Co-design, with sales, generalized training that reflects new tactics and attitudes about our customers, 1Q 20XX. (R.P.)

b. Work closely with marketing campaign team to design suitable training for new programs, goals TBD.

Potential Cross-Team Issues

Continue to nurture closer working relationship between marketing and sales: form joint implementation team?

Need support from training department on customer care training design and delivery.

Need to work with personnel on hiring criteria to reflect new attitudes?

B. Internal challenge: increase the success of our sales campaigns, measured against rate of sales conversions of responders (using 1Q 1998 campaign as a baseline) with specific goal TBD based on campaign targets set by marketing team. (R.Ps.)

Guiding Assumptions

We believe the weaknesses in our current sales campaigns are the result of two general factors: ineffective campaign execution and the need for a revised campaign approach.

Specifically, we want to improve the following: (1) the breakdown of internal communications between marketing and sales; (2) inadequate follow-up after the initial customer contact has been made; (3) a "one-size-fits-all" approach to the customer (note crossover with overall marketing goal A above).

Strategic Actions/Tasks

1. Marketing/sales joint task force: establish a marketing and sales support team by 1Q 20XX—including sales, customer service, fulfillment—with the following objectives: (R.P.)
   a. Review current hand-off practices for sales campaigns, 1Q 20XX. (R.P.)
   b. Consider and/or design special program for new customers: welcome call and survey, and periodic check-in mailers 3Q 20XX. (R.P.)
   c. Design and delivery of training for internal personnel who handle new accounts 2Q 20XX. (R.P.)
2. Email tech team tasks: support/initiate project for the enhancement of headquarters-to-branch email capacity, Jan 15, 20XX. (R.Ps.)
3. Web site task force task: assess computer sophistication of current customers and, if warranted, meet with IT team to discuss the marketing and distribution potential for extranet Web site, 2Q 20XX. (R.Ps.)

Potential Cross-Team Issues

IT and Marketing/Sales initiatives will require cross-team support.

IT may also need to be involved in current customer computer-use analysis.

Will there be shared budget issues on potential extranet initiatives?

C. Department challenge: reduce paper use in the department by 15% and copier usage by 10% by 3Q 20XX (R.P.) and expedite new contract account set-up from 8 weeks to 3, by end-of-year 20XX. (R.Ps.)

Guiding Assumptions

The assumption is that increased paper and copier usage can be combatted by environmental education and a focus on environmentally-friendly practices in the department.

One also assumes the Legal Department will work with us to consider and improve the efficacy of contracting procedures.

Strategic Actions/Tasks

1. Contracting procedures task force: in collaboration with Legal, review contracting procedures in order to reduce the amount of time for new contract account set-up from eight weeks to three, by end-of-year 20XX. (R.P.)
2. Reduce paper use task force:
    a. Open broader discussion with department members about preferred communications modes: paper, electronic/digital media, face-to-face, etc. At next department meeting. (R.P.)
    b. Suggest at next quarterly division head meeting that a task force be appointed and a series of environmentally-friendly policies be developed, 1Q 20XX. (R.P.)
    c. Research national or local environmental organizations to see what material already exists, 2Q 20XX. (R.P.)
    d. Research broader applications of an environmental policy: copier cartridges, laser-jet ink replacements, recycling of all office waste materials, etc., 2Q 20XX. (R.P.)
    e. Disseminate information and begin campaign by 3Q 20XX. (Wrap decreased copier and paper usage goals inside the corporate campaign.) (R.P.)

Potential Cross-Team Issues

To be successful, the environmentally-friendly policies must be adopted corporate-wide.

6. ADDING MONITORING TOOLS

> **A. Increase our total market share by 8% in 2 years, to be tracked quarterly, by increasing sales in our existing three-state region by roughly 1% per quarter beginning 2Q 20XX. (R.P.) (Base measurement: total market share as of year-end 20XX.)**
>
> *Monitoring:* as specified above will be computed and compiled by R.P. and passed on to R.P. for the following distribution:
>
> ■ Intranet: "Planning Goals," posted the last month of every quarter. (R.P.)
> ■ Meeting notes: included in a special tracking section of division managers meeting notes. (R.P.)
> ■ Email: posted via email to division heads as soon as available. (R.P.)
> ■ Email: to all project team members. (R.P.)

Guiding Assumptions

Our core competencies focus on people skills and strong branding, not technology or low-price leadership.

Therefore, growth will be achieved by deepening relationships with current customers and by partnering with them to gather intelligence in order to increase our customer base.

Major thrust will be a renewed customer service orientation in preparation for potential new product design and launch (beginning 4Q 20XX).

No dramatic monetary offers will be utilized at this time, except the perceived value of courtesy and closer personal contact.

Strategic Actions/Tasks

1. Customer intelligence survey team tasks include the following:
   a. Identify survey research and design team by 4Q 20XX. (R.P.)
   b. Review existing marketing data on the following items by 1Q 20XX: (R.P.)
      - Customer preferences for current products: sales profile
      - Customer suggestions for new products
      - Reasons for customers' choosing Competitor X's product
   c. If existing data is insufficient, design survey vehicle and implementation plan to capture needed marketing data by 2Q 20XX. (R.P.)

    d. Assemble production team to research printing and distribution mechanics to have survey in customers' hands by mid-1Q 20XX. (R.P.)

    e. Develop analysis tools to evaluate our current customers and understanding them using finer demographic customer profiles, based on survey results, 2Q 20XX. (R.Ps.)

    f. Follow up with 'We Need You' calls for portion of customers who do not return surveys by end 3Q 20XX. (R.P.)

    g. Research the feasibility of implementing customer surveys for 5% of total current customer base on a quarterly basis. With a go/no-go answer by 3Q 20XX. (R.P.)

    h. Hire mystery shoppers to shop in areas of projected expansion and reporting back to marketing implementation team by 2Q 20XX. (R.P.)

2. Marketing campaign redesign and execution team tasks include the following:

    a. Develop 3Q marketing campaign with a diverse array of approaches—including initial contact vehicle and series of follow-up tools—to match unique, customer-specific demographic niches. (R.Ps.)

    b. Research utilization of Web site for partnering program using permission marketing techniques, beginning 2Q 20XX. (R.P.)

    c. Consider new series of testimonial ads in target market area, mid-2Q 20XX. (R.P.)

    d. Pursue partnership with our retail outlets to institute "Open House" tables and demonstrations at local outlets with the purpose of establishing personal contact with our customer base, mid-3Q of end 4Q 20XX. (R.P.)

    e. Consider the formation of "You Tell Us How" teams of consumers to advise marketing group, in exchange for special customer offers, "Thank You" gatherings, and appreciation awards, 2Q 20XX. (R.P.)

3. Training design and delivery team tasks include the following:

    a. Co-design, with sales, generalized training that reflects new tactics and attitudes about our customers, 1Q 20XX. (R.P.)

    b. Work closely with marketing campaign team to design suitable training for new programs, goals TBD.

Potential Cross-Team Issues

Continue to nurture closer working relationship between marketing and sales: form joint implementation team?

Need support from training department on customer care training design and delivery?

Need to work with personnel on hiring criteria to reflect new attitudes?

B. Internal challenge: increase the success of our sales campaigns, measured against rate of sales conversions of responders (using 1Q 1998 campaign as a baseline) with specific goal TBD based on campaign targets set by marketing team. (R.Ps.)

Monitoring: responder rate data will be computed and compiled by R.P. daily during both the pre-holiday and spring campaigns and distributed by R.P. as follows:

- Marketing bulletin board by the copy machine by 9:00 AM for previous day. (R.P.)
- Email posting to all marketing/sales team members by 10:00 AM for the previous day. (R.P.)
- Aggregated and reported at weekly department meetings for marketing, sales, customer service and fulfillment. (R.Ps.)
- Aggregated and reported in bi-weekly status reports to division heads. (R.P.)
- Final results included in quarterly status report to Board. (R.P.)

Guiding Assumptions

We believe the weaknesses in the current sales campaigns are the result of two general factors: ineffective campaign execution and the need for a revised campaign approach.

Specifically, we want to improve the following: (1) the breakdown of internal communications between marketing and sales; (2) inadequate follow-up after the initial customer contact has been made; and (3) a "one-size-fits-all" approach to the customer (note crossover with overall marketing goal A above).

Strategic Actions/Tasks

1. Marketing/sales joint task force: establish a marketing and sales support team by 1Q 20XX—including sales, customer service, fulfillment—with the following objectives: (R.P.)

   a. Review current hand-off practices for sales campaigns, 1Q 20XX. (R.P.)

      b. Consider and/or design special program for new customers: welcome call and survey, and periodic check-in mailers, 3Q 20XX. (R.P.)

      c. Design and delivery of training for internal personnel who handle new accounts, 2Q 20XX. (R.P.)

  2. Email tech team tasks: support/initiate project for the enhancement of headquarters-to-branch email capacity, Jan 15, 20XX. (R.Ps.)

  3. Web site task force task: assess computer sophistication of our current customers and, if warranted, meet with IT team to discuss the marketing and distribution potential for extranet Web site, 2Q 20XX. (R.Ps.)

Potential Cross-Team Issues

IT and Marketing/Sales initiatives will require cross-team support.

IT may also need to be involved in current customer computer-use analysis.

Will there be shared budget issues on potential extranet initiatives?

C. Department challenge: reduce paper use in the department by 15% and copier usage by 10% by 3Q 20XX (R.P.) and expedite new contract account set-up from 8 weeks to 3, by end-of-year 20XX. (R.Ps.)

Monitoring: track paper and copier usage by month and report as follows: (R.P.)

- Marketing bulletin board by the copy machine by the first Tuesday following month-end. (R.P.)
- Aggregated and reported in bi-weekly status reports to division heads. (R.P.)

Monitoring: progress on new contracting procedures will be documented in weekly status reports by R.P. to department heads.

Guiding Assumptions

The assumption is that increased paper and copier usage can be combatted by environmental education and a focus on environmentally-friendly practices in the department.

It is assumed that legal will work with us to consider and improve the efficacy of contracting procedures.

Strategic Actions/Tasks

1. Contracting procedures task force: in collaboration with legal, review contracting procedures in order to reduce the amount of time for new contract account set-up from eight weeks to three, by end-of-year 20XX. (R.P.)
2. Reduce paper use task force:
   a. Open broader discussion with department members about our preferred communications modes: paper, electronic/digital media, face-to-face, etc. At next department meeting. (R.P.)
   b. Suggest at next quarterly division head meeting that a task force be appointed and a series of environmentally-friendly policies be developed, 1Q 20XX. (R.P.)
   c. Research national or local environmental organizations to see what material already exists, 2Q 20XX. (R.P.)
   d. Research broader applications of an environmental policy: copier cartridges, laser-jet ink replacements, recycling of all office waste materials, etc., 2Q 20XX. (R.P.)
   e. Disseminate information and begin campaign by 3Q 20XX. (Wrap decreased copier and paper usage goals inside the corporate campaign.) (R.P.)

Potential Cross-Team Issues

To be successful, the environmentally-friendly policies must be adopted corporate-wide.

GLOSSARY

agreed to: one of the attributes of an effective goal; meaning reached by consensus.

assignable: one of the attributes of an effective goal; meaning that there is a group of people accountable for its accomplishment.

audience: in the context of monitoring, the person or persons to whom the statistical information will be reported.

challenge: identifying challenges is the first step in our planning methodology; meaning something that "needs fixing," an opportunity or difficulty that the business is facing.

cluster: a group of challenges that share the same root challenge or are part of the same system or family of problems.

clustering: the process of grouping like ideas in systems or families.

collaboration: a group of people working together on their own independent parts of a project; they are dependent on one another for a successful outcome, but they have their own autonomy and areas of authority.

collateral material: the commonly used term to refer to all marketing material used for external customer communications, for example, business cards, letterhead, envelopes, marketing brochures, etc.

control: generally, some standard or measurement used to regulate or verify the results of an experiment; or, in the case of planning, a means of measuring the success of an action or activity.

core competencies: the particular or essential strengths of an organization, for example, fund raising, R & D, customer service, strong branding, etc.

corporate culture: a system of beliefs and values shared by members of a given company.

flexible: one of the attributes of an effective goal; meaning that the goal and/or the means of accomplishing the goal can be easily adapted to changing business conditions.

fuzzy goals: goals that are not easily quantifiable; generally non-financially-based goals common in the areas of customer service, PR, employee relations, etc.; for example, "We want to raise customer satisfaction," or "We will strive to improve our image as a corporate good citizen within our target market"; as opposed to a concrete goal such as "We will increase our market share by 4% next year."

goal or objective: the second step in our planning methodology; meaning something one strives for, something one puts mental attention on in order to achieve.

guiding assumptions: the conditions underlying a particular business or organizational environment; parameters that define, limit, or direct a particular strategic action.

insight: the ability to see the true nature of a situation, event, or person by perceiving what is essential or inherent *before* an outcome or event makes it apparent.

inspirational: one of the attributes of an effective goal; meaning that the goal reflects the values and mission of the organization, and, further, that team members believe it is worthy of their best efforts.

KC: key competencies; what a business does best.

KRF: key result factors; a.k.a. goals.

market share: a company's share of the market, measured as a proportion either in the amount of the company's revenue to the total revenue in the marketplace; or units sold to the total possible number of units sold for a particular product.

MBO: management by objectives; a way to create accountability for specific corporate goals by linking them to actions that are built into compensation or job performance requirements.

MBWA: literally, management by walking around; a management style that advocates making personal contact by walking the floor of a business.

measurement unit: in monitoring, how the goal measurement will be quantified and tracked; in a simple example, by monitoring copier usage based on number of copies made as recorded in the copy machine.

measurer: the person responsible for gathering monitoring statistics; the statistician who gathers tracking information for strategic action goals.

measuring frequency: in monitoring, how often the measurement will be taken: for example, collecting copier usage information weekly, at the close of business on Friday.

mind map: a free-associative sketch of ideas related to a particular theme.

monitoring and measurement: the name for the fourth step of our planning methodology; meaning the process of quantifying a goal, assigning responsibility for its completion, and creating a system for its measurement and tracking.

perceived value: the value *to the customer* of a marketing gift, giveaway, or prize; often, the perceived value of an item is far greater than its actual value.

permission marketing: a term created and used by Seth Godin that outlines a consensual relationship model that establishes a valid reason for a customer to give you information in exchange for something of value to him or her, (e.g., a prize or a gaming situation).

purpose of strategic action planning: to translate broad management objectives into detailed action plans to ensure selected business results.

quantifiable: one of the attributes of an effective goal; meaning that the achievement of the goal can be statistically measured.

rate of sales conversion of responders: the proportion of the number of customers who respond to an offer and can be converted into buyers to the total number of all customers who respond; for example, one expects a certain percentage of customers to respond to a direct mail offering by calling or sending in the coupon to register for a contest, or whatever; it is the percentage of these responders who are convinced to actually buy the product or open the checking account, or whatever, that is referred to.

reachable: one of the attributes of an effective goal; meaning not so unrealistic that it is demoralizing, but enough of a stretch to be a challenge; doable.

relative market share: a company's market share relative to the position of competitors within a particular market.

relevant: one of the attributes of an effective goal; meaning that the goal is aligned with the corporate mission and strategic direction of the company and appropriate for the people who will be assigned to do it.

report vehicle: the actual communication device for delivering the monitoring statistic; for example, email, memo, annual report, company newsletter, etc.

reporting format: in monitoring, the particular physical form for communicating a set of statistics: for example, the format could be a percentage used in a prose report, a pie-chart graphic, or a spreadsheet.

reporting frequency: in monitoring, the intervals at which certain statistical information will be reported; for example, in a quarterly report. Note that the measuring frequency might be weekly, although the reporting frequency for certain audiences could be quarterly.

ROI: return on investment.

root challenge: a common cause for several challenges that may have been identified separately; if two or more challenges share a root challenge, they should be included in the same cluster.

sales profile: historical data about a customer's buying preferences: what products have been purchased and when.

SME: subject matter expert; the term for anyone who has insider or special technical knowledge that might be needed by someone else in order to complete a job.

strategic action: the third step of our planning methodology; meaning a targeted activity that optimizes efforts and resources to produce selected business results.

strategy: a general concept that governs a series of specific actions or provides clear and imaginative guidance for the effective use of available resources in order to reach a specific goal.

SWOT: strengths, weaknesses, opportunities, and threats; a situational analysis tool.

team: individuals who have their group process and project result as their top priority.

Total Quality Management (TQM): a systematic process created by W. Edwards Deming for controlling and managing error in business production and manufacturing.

trends and fads: tendencies of large groups of people toward certain types of behaviors and activities.

tshatshki: a Yiddish word meaning toy or doo-dad.

understandable: one of the attributes of an effective goal; meaning clear and commonly understood among implementation team members.

visible: one of the characteristics of an effective goal; meaning that the tracking for and progress being made against the goal is apparent.

wordsmith: someone who is expert in the craft of wordsmithing, i.e., writing and crafting ideas using the tool of language.

wordsmithing: the process of crafting the final phrasing for a document or idea.

x-y axis quadrant model: a visual analysis tool that uses two variables plotted on the x- and y-axes to produce a cube with four quadrants which allows the identification of sweet-spots or target regions depending on the variables used; for example, see Impact vs. Resources Cube, Figure 6.2.

SELECTED READINGS

These are the books that I found most helpful or inspirational in the process of writing this book. They deliver clear information on specific topics and speak to planning from a variety of different angles, from very formal to quite casual, from very practical to almost spiritual.

OVERALL TEAM-BASED PLANNING PROCESS

Fogg, C. Davis, *Team-Based Strategic Planning*, Amacom, New York, 1994. Provides a good overview of the planning process with a focus on working with a team. The approach is very people-oriented, using real-world examples. Very detailed meeting agendas, time-tables, and sample forms. Back matter includes samples of two complete strategic plans.

Barry, Bryan W., *Strategic Planning Workbook for Nonprofit Organizations*, Amherst H. Wilder Foundation, St. Paul, MN, 1986. This is a useful text that clearly outlines a strategic planning procedure for the non-business professional. It is a large format workbook (8½" by 11") with jolly illustrations, a friendly layout, and straightforward language. A five-step process is outlined—Get organized, Take stock, Develop a strategy, Draft and refine the plan, and Implement the plan. The chapter headings include: "Introduction to Strategic Planning," "Developing Your Strategic Plan," and "Appendices." That's it! There are lots of worksheets in the Appendices meant to be copied. Although the focus is for the nonprofit organization, the approach is applicable to many business situations.

GENERAL BUSINESS CONCEPTS

Argenti, Paul A., *The Fast Forward MBA*, John Wiley & Sons, New York, 1997. There is an especially good and concise discussion of trends in management and strategic planning models in the chapter on "Strategic Management." It includes a quick review of the thinking of Alfred Chandler, Frederick Taylor, W. Edwards Deming, Michael Porter, the McKinsey 'Seven S' model and Bruce Henderson's Boston

Consulting Group 'star-cow-dog-?' model, and some recent research from the Harvard Business School. The book is also an invaluable resource for general business concepts.

STRATEGIC THINKING

Ohmae, Kenichi, *The Mind of the Strategist—The Art of Japanese Business*, McGraw-Hill, New York, 1982. A classic discussion of strategy—what it is, how to enhance your strategic skill, and how to apply it to particular business situations.

Bandrowski, James F., *Corporate Imagination Plus*, The Free Press (Collier Macmillan), Berkeley, CA, 1990. This book discusses what is needed to make the strategic planning process responsive and creative. Some representative chapter titles indicate the focus of the book: "Developing Strategic Insights," "Taking Creative Leaps," "Making Strategic Connections." Bandrowski's approach to planning is creative—for example, there is an essay in the Appendix on "The Nature of Genius" that compares, in a pop-psychology way, similarities between the creativity of CEOs and scientific thinkers like Einstein, Darwin, and Röntgen (the discoverer of X-rays). Lots of good ideas and packed with business vignettes and anecdotes.

Peters, Thomas J. and Robert H. Waterman, Jr., *In Search of Excellence: Lessons from America's Best Run Companies*, Warner, New York, 1982. The number one best seller in 1983 and still relevant. It describes the eight basic principles that the best-run companies utilize to keep on top.

VISUAL TOOLS FOR TEAM MEETINGS AND PRESENTATIONS

Richey, Terry, *The Marketer's Visual Tool Kit: Using Charts, Graphs and Models for Strategic Planning and Problem Solving*, American Management Association, New York, 1994. Contains an intriguing description of how the triangle, circle, square, and other shapes can be used in brainstorming. Richey outlines an imaginative and highly usable approach for sparking creativity and strategic thinking in planning sessions. The graphics are also instantly accessible for presentation and report purposes. Richey proposes formats for intuitive visual concepts, not complex statistical graphs.

MISSION/VISION AND ALIGNMENT

Albrecht, Karl, *The Northbound Train: Finding the Purpose, Setting the Direction, Shaping the Destiny of Your Organization*, American Management Association, New York, 1994. An overview of strategic planning from the perspective of everything that must be in place before planning can begin: leadership characteristics of the CEO; how to create a corporate vision; creating and aligning behind a mission statement; core values; business modeling; and customer-focused initiatives. Albrecht has an excellent section called "Vision, Mission, Values," in which he presents several actual corporate vision and mission statements and critiques them.

Abrahams, Jeffrey, *The Mission Statement Book: 301 Corporate Mission Statements from America's Top Companies*, Ten Speed Press, Berkeley, CA, 1995. If your company is considering making or redoing your mission or vision statement, this book might be worthwhile reading. It includes a compilation of mission statements from over 300 companies. It is interesting to read through the book and look at the variety of different forms and approaches businesses have taken to encapsulate their beliefs.

SITUATIONAL ANALYSIS AND ECONOMIC MODELS

Marrus, Stephanie K., *Building the Strategic Plan: Find, Analyze and Present the Right Information*, John Wiley & Sons, New York, 1984. This book outlines a classic MBA-style approach to situational analysis, including complete descriptions of share-matrix, share momentum, profitability models, and definitions of terms, etc. The author provides explanations of analysis models, and how to find and analyze the data so that it can be graphically represented and compared easily. The first two chapters also give a good overview and explanation of what elements should be present in a corporate plan. The last chapter in the book reviews some analyst firms and gives suggestions about how to work with them if you want to buy a comprehensive situational analysis rather than conduct one yourself. Note that the book is almost 15 years old, so the analyst information is dated—however, the situational analysis approach is timeless.

Porter, Michael E., *Competitive Strategy: Techniques for Analyzing Industries and Companies*, Free Press, New York, 1980. By now a classic text that outlines a competitive analysis model, including five key components for consideration: competitor's response profile, future goals, assumptions, current strategy, and capabilities.

PHILOSOPHY OF PLANNING

Steiner, George A., *Strategic Planning: What Every Manager Must Know*, The Free Press (Collier Macmillan), 1979. Steiner is the legendary master of traditional strategic planning. This book analyzes different methods for establishing a corporate planning process. There is a lot of discussion about buy-in, styles of management, how to make the process effective, and whether to use a bottom-up or top-down approach. Steiner discusses the many different kinds of planning processes, sometimes giving the history and evolution of each. Packed with information, this book presents an in-depth and almost historical study of planning. A classic.

GENERAL INSPIRATION

Whyte, David, *The Heart Aroused: Poetry and the Preservation of the Soul in Corporate America*, Doubleday, New York, 1994. If you are just tired of the grind and want a totally new perspective on the corporate world, take at look at Whyte's book. Whyte is a poet and business consultant who weaves poetry, storytelling, and mythic archetypes into his comments about life in the modern corporation. A real find. This book could change your life.

BIBLIOGRAPHY

Abrahams, Jeffrey, *The Mission Statement Book: 301 Corporate Mission Statements from America's Top Companies*, Ten Speed Press, Berkeley, CA, 1995.

Ackoff, Russel L., *A Concept of Corporate Planning*, Wiley-Interscience, New York, 1971.

Albrecht, Karl, *The Northbound Train: Finding the Purpose, Setting the Direction, Shaping the Destiny of Your Organization*, American Management Association, New York, 1994.

Argenti, Paul A., *The Fast Forward MBA*, John Wiley & Sons, New York, 1997.

Axelrod, Robert, *The Evolution of Cooperation*, BasicBooks, HarperCollins, New York, 1984.

Bandrowski, James F., *Corporate Imagination Plus*, The Free Press (Collier Macmillan), New York, 1990.

Barry, Bryan W., *Strategic Planning Workbook for Nonprofit Organizations*, Amherst H. Wilder Foundation, St. Paul, MN, 1986.

Below, Patrick J., *The Executive Guide to Strategic Planning*, Jossey-Bass, San Francisco, 1987.

Bronowski, Jacob, *The Origins of Knowledge and Imagination*, Yale University Press, New Haven, CT, 1978.

Drucker, Peter F., *Management: Tasks, Responsibilities, Practices*, Perennial Library, Harper & Row, New York, 1974.

Drucker, Peter F., *Managing in a Time of Great Change*, Truman Talley Books/Dutton, New York, 1995.

Everdell, William R., *The First Moderns: Profiles in the Origins of Twentieth-Century Thought*, University of Chicago Press, Chicago, 1997.

Fogg, C. Davis, *Team-Based Strategic Planning, A Complete Guide to Structuring, Facilitating and Implementing the Process*, American Management Association, New York, 1994.

Hamilton, Edith, *Mythology: Timeless Tales of Gods and Heroes*, The New American Library, New York, 1942.

Lathem, Edward Conney, In the Home Stretch, *The Poetry of Robert Frost*, Holt, Rinehart and Winston, New York, 1969.

Lorange, Peter, and Richard F. Vancill, *Strategic Planning Systems*, Prentice-Hall, Englewood Cliffs, NJ, 1977.

Marrus, Stephanie K., *Building the Strategic Plan: Find, Analyze and Present the Right Information,* John Wiley & Sons, New York, 1984.

Mason, Stanley, *Inventing Small Products, for Big Profits Quickly,* Crisp Management Library, Menlo Park, CA, 1997.

Ohmae, Kenichi, *The Mind of the Strategist: The Art of Japanese Business,* McGraw-Hill, New York, 1982.

Peters, Thomas J., and Robert H. Waterman, Jr., *In Search of Excellence,* Time Warner Books, New York, 1982.

Popcorn, Faith, *The Popcorn Report,* HarperBusiness, New York, 1992.

Porter, Michael E., *Competitive Strategy: Techniques for Analyzing Industries and Companies,* Free Press, New York, 1980.

Powers, Charles W., *Ethics in the Education of Business Managers,* John Wiley & Sons, New York, 1991.

Richey, Terry, *The Marketer's Visual Tool Kit: Using Charts, Graphs and Models for Strategic Planning and Problem Solving,* American Management Association, New York, 1994.

Sardello, Robert, *Love and the Soul:* Creating a Future for Earth, HarperCollins, New York, 1995.

Schrage, Michael, *No More Teams! Mastering the Dynamics of Creative Collaboration,* Currency, Doubleday, New York, 1995.

Simons, George, *Working Together,* Crisp Publications, Menlo Park, CA, 1989.

Steiner, George A., *Strategic Planning: What Every Manager Must Know,* The Free Press (Collier Macmillan), 1979.

Tannen, Deborah, *Talking from 9 to 5: How Men's and Women's Conversational Styles Affect Who Gets Heard, Who Gets Credit and What Gets Done at Work,* William Morrow and Company, New York, 1994.

Toffler, Barbara Ley, *Tough Choices: Managers Talk Ethics Making Tough Choices in a Competitive Business World,* John Wiley & Sons, New York, 1991.

Verger, Morris D. and Norman Kaderlan, *Connective Planning,* McGraw-Hill, New York, 1993.

Wheatley, Margaret J. and Myron Kellner-Rogers, *A Simpler Way,* Berrett-Koehler, San Francisco, 1996.

Whyte, David, *The Heart Aroused: Poetry and the Preservation of the Soul in Corporate America,* Doubleday, New York, 1994.

Wildavsky, Aaron, Does Planning Work, *Public Interest,* Summer 1971.

Wilson, Edward O., *Biophilia: The Human Bond with Other Species,* Harvard University Press, Cambridge, MA, 1984

Wind, Jerry Yoram, and Jeremy Main, *Driving Change: How the Best Companies Are Preparing for the 21st Century,* The Free Press, New York, 1998.

INDEX